This book contains materials developed by the AIMS Education Foundation. **AIMS** (**A**ctivities **I**ntegrating **M**athematics and **S**cience) began in 1981 with a grant from the National Science Foundation. The non-profit AIMS Education Foundation publishes hands-on instructional materials (books and the monthly magazine) that integrate curricular disciplines such as mathematics, science, language arts, and social studies. The Foundation sponsors a national program of professional development through which educators may gain both an understanding of the AIMS philosophy and expertise in teaching by integrated, hands-on methods.

ISBN **1-881431-73-8**

Printed in the United States of America

Spatial Visualization

❦ ❦ ❦ ❦ ❦ ❦ ❦ ❦ ❦ ❦ ❦ ❦ ❦ ❦ ❦ ❦ ❦ ❦

Author
Arthur J. Wiebe

Editor
Betty Cordel

Illustrator
Brenda Richmond

Desktop Publisher
Roxanne Williams

*The **AIMS Education Foundation** is a research and development organization dedicated to the improvement of the teaching and learning of mathematics and science through a meaningful integrated approach.*

Table of Contents

A Model of Mathematics

Spatial Visualization
Introduction

"Real-world" geometry is the geometry of greatest value to students. Real-world geometry consists of studies in three dimensions, including solids and their plane derivatives. The preparation of students to utilize geometry in the real world necessitates extensive experience exploring solids, their properties and interrelationships in an environment that nurtures understanding. An essential characteristic of such an environment is that it provides for frequent and varied opportunities to study the three-dimensional aspects of geometry. The AIMS *Studies in Geometry* series, which includes this publication, incorporates such an emphasis.

There is good reason why the national reform documents and state mathematics frameworks place emphasis on the development of spatial visualization skills. Those skills are essential for working successfully with solids in real-world applications. Research reports a strong correlation between spatial visualization and problem-solving ability and the significant contribution this makes to the total learning process. Of special interest is that studies indicate **spatial visualization instruction benefits girls and boys similarly.**

Students develop spatial concepts and relationships best through activities in which they manipulate and carefully observe concrete objects. Since the development of spatial skills takes time and patience, it is incumbent upon all mathematics educators to allocate sufficient time and meaningful treatment to this aspect of the curriculum. Students need to develop understandings in a nurturing environment that provides differentiated tasks meeting their individualized needs.

As students acquire facility in spatial visualization, they come to realize that the study of geometry is both fascinating and of obvious value in the real world for which they are preparing. It is a crucial skill for engineers, architects, builders, astronauts, artists, and hundreds of other professionals. Students must combine experiences manipulating and analyzing solids with translations between two-dimensional and three-dimensional representations and models. The ability to visualize and construct three-dimensional models from two-dimensional plans and to represent three-dimensional models in two dimensions are essential components in mathematics and science education.

Spatial Visualization provides a range of challenging explorations through which students can develop and hone their spatial visualization skills. Building solids from the four-view plans and isometric drawings, reducing solids to four-view plans and isometric drawings, creating "exploded" views of the layers in a solid, studying maximum and minimum volumes and surface areas, analyzing isometric drawings to determine volumes, surface area, and perimeters of the footprints of solids, and exploring polycubes exemplify the range found in this publication.

The AIMS *Model of Mathematics* and its derivative, the *Five-Star Teaching/Learning Environment*, serve as meaningful references for understanding the four environments in which mathematics is experienced and the interrelationships of the processes utilized in these activities. The four distinct environments depicted in the models are:

- experiences in the real world (circle),
- writing expressions using the symbols of mathematics and language (triangle),
- illustrating the real world by using drawings, graphs, and other types of picturing (square), and,
- thoughtful analysis utilizing hypothesizing, drawing conclusions, making generalizations, etc. (hexagon).

The ability to translate from one environment to another is crucial.

Greatest emphasis is placed on real-world geometry experiences with physical models (circle) and their representational counterparts (square). When students quantify perimeters, surface areas, volumes, and the number of vertices, they exemplify working in the triangle environment. As they problem solve, search for all solutions, and discover principles, etc., they operate in the hexagon environment. While the environments denoted by the circle and square will predominate, all four are always involved.

The Five Star Model

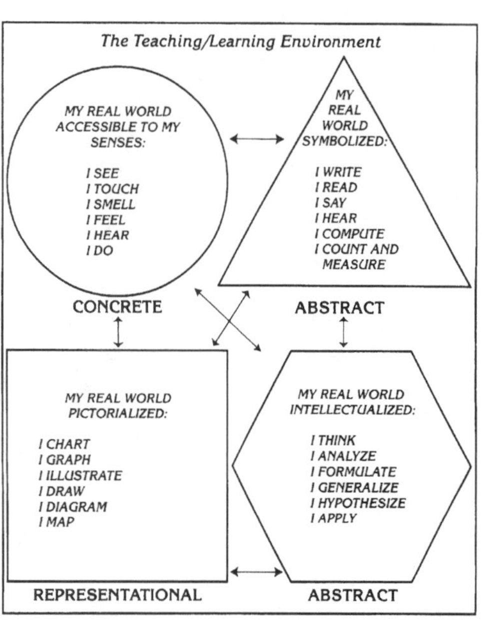

Content of *Spatial Visualization* and its Relationship to the *Model Of Mathematics*

In *Part One*, students build models from four-view plans which show top, front, left- and right-end views. Referring to their constructed model, they draw a back view and determine the perimeter, surface area, volume, and number of vertices.

In *Part Two*, students refer to an isometric drawing in constructing a geometric solid. They draw four-view plans by referring to the model they have constructed. Students continue to determine the perimeter, surface area, and volume of each figure.

Students are introduced to making isometric drawings in *Part Three*. This is an essential skill of great value that they will utilize extensively in subsequent activities and real-world applications. Students will sense a surge of power and interest as they develop a feel for making isometric drawings. Each activity begins by constructing a solid from four-view plans followed by making an isometric drawing. Volumes, surface areas, and perimeters continue to be determined.

Part Four introduces the drawing of exploded isometric views. *This skill can be taught at an any point deemed appropriate.* Drawing exploded isometric views represents a major advance in difficulty and the ability to visualize spatially. However, its mastery provides a powerful impetus for the further development of spatial visualization capabilities. Students who persevere will find the subsequent challenges interesting and fascinating.

Assignments to draw exploded isometric views can be associated with any of the activities throughout this publication. It is recommended that students go back to the beginning and work with the easier models in *Part One* and proceed progressively through the more complex models found in later sections. In *Part Four* students construct models based on four-view plans and then make isometric and exploded isometric drawings. They continue to determine base perimeters, surface areas, and volumes.

In *Part Five* students are presented with isometric drawings of solids nestled into the corner of a box. The hidden sides are smooth. Students are challenged to determine the volume, surface area, and base perimeter by studying the drawings. They conclude by making exploded view drawings.

Maximum-minimum and other problem-solving activities constitute the content of *Part Six*. Solids are constructed using a given number of cubes to achieve a minimum or maximum surface area, students construct solids of their own design with a given volume and surface area, and longer-term investigations with pentominoes and pentacubes are suggested. These explorations emphasize congruency and similarity. Isometric drawings are associated with each activity.

Spatial Visualization

Part One: Building Models from Four-View Plans

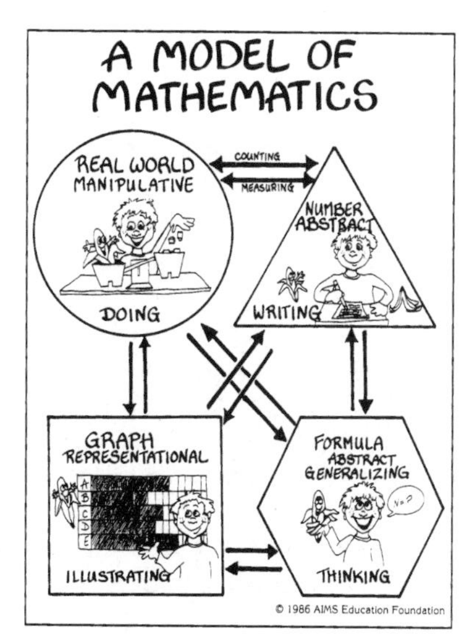

In each of the five introductory activities, students are provided plans which embody four views of solids they are to build: top, left, front, and right. In the AIMS *Model of Mathematics* this represents the movement from the square (a two-dimensional picture or plan) to the circle (a three-dimensional structure). After the solid is built, students are asked to draw the back view using the scheme as in the other views. Going from the solid to the plan introduces the inverse of going from the plans to the solid. Such reversals in procedure possess high instructional value.

The key to interpreting the four views is to understand that a bold black line separating cubes indicates a break in the surface. The surface of a block on one side of the bold black line is nearer to the observer than the surface on the other side. Any surface with only narrow black lines separating cubes is smooth, while any surface where bold black lines separate cubes is not smooth.

The introductory structures in this section are simple enough so most students will experience early success while learning to "read" the plans. It is important to have such successful experiences to build confidence. The difficulty level increases gradually as students progress through the activities.

The Interlocking Cubes

While any type and size of interlocking cubes may be used, the AIMS *Hex-a-Link Cubes* are especially useful for activities in this and several other AIMS publications. These cubes offer greater flexibility now that AIMS has added inserts. These added inserts are for use in those instances where more than one protrusion per cube is needed to provide the desired stability in certain structures as in skeletal outlines of solids. Because these blocks are used so extensively in a number of studies in geometry publications, their acquisition is highly cost effective.

Open-Ended Activities to be Encouraged

As soon as students gain an understanding of all that is involved in building structures from four-view plans, they should be encouraged to design their own structures and draw the related four-view plans. A recommended approach is to have each student design and build a structure, draw the four-view plans, conceal the structure, and exchange plans with another student who has done the same. Each student then builds the structure according to the plans that have been received. After the structures are completed, students compare them with the originals from which the plans were drawn. If they match, then the steps in the process (structures-plans-exchange of plans-structures) have been done correctly. If they do not match, students must analyze each step in the process to determine where an error occurred. This creative involvement is both motivational and instructive.

Measures

To standardize the measures for all types of interlocking cubes, **the measure of one edge is defined as one unit of length, the measure of the area of one face as one square unit, and the volume of one cube as one cubic unit.** Use of these units makes it possible to use interlocking blocks of any size.

Vertices Defined

A vertex may be either on an outside or inside "corner" as indicated in the illustration. In three-dimensional figures, students often think of vertices as only those that feel like a point when touched. In these structures, the inside corners cannot be touched in the same way but must still be counted as vertices. The figure in the

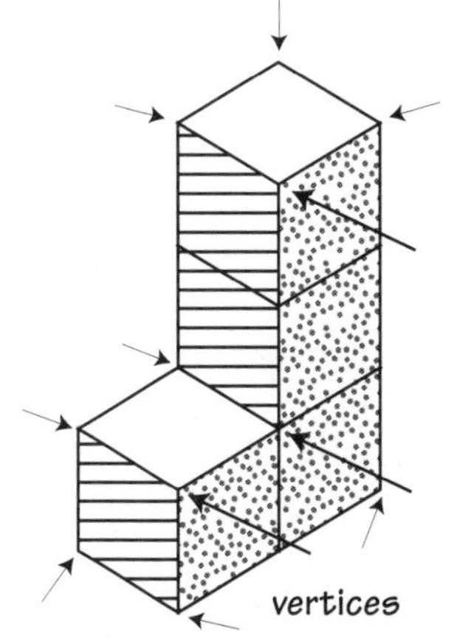

illustration has 12 vertices, 11 of which are indicated by arrows. The twelfth is on the far corner of the base and is not visible.

After each structure is built, students are asked to find the volume, total surface area, number of vertices, and the greatest and least perimeters among the four views.

Answers

Thinkcard 1

1. Volume: 4 units^3
2. Least perimeter: 6 units, top view
   Greatest perimeter: 10 units, front view
3. Total surface area: 18 units^2
4. Number of vertices: 12

Thinkcard 2

1. Volume: 5 units^3
2. Least perimeter: 8 units, top, left, right views
   Greatest perimeter: 12 units, front view
3. Total surface area: 12 units^2
4. Number of vertices: 16

Thinkcard 3

1. Volume: 5 units^3
2. Least perimeter: 8 units, left, right views
   Greatest perimeter: 10 units, top, front views
3. Total surface area: 22 units^2
4. Number of vertices: 21

Thinkcard 4

1. Volume: 5 units^3
2. Least perimeter: 8 units, top, left, right views
   Greatest perimeter: 12 units, front view
3. Total surface area: 24 units^2
4. Number of vertices: 16

Thinkcard 5

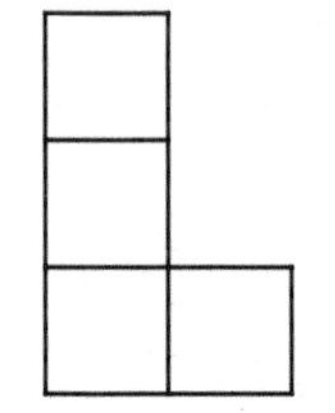

1. Volume: 6 units^3
2. Least perimeter: 8 units, top view
   Greatest perimeter: 10 units, left, front, right views
3. Total surface area: 24 units^2
4. Number of vertices: 14

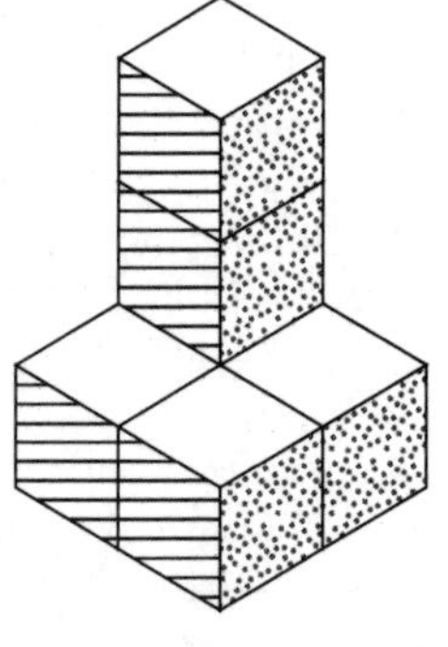

Please construct the figure whose plans are shown. Consider the length of the edge of a block to measure one unit.

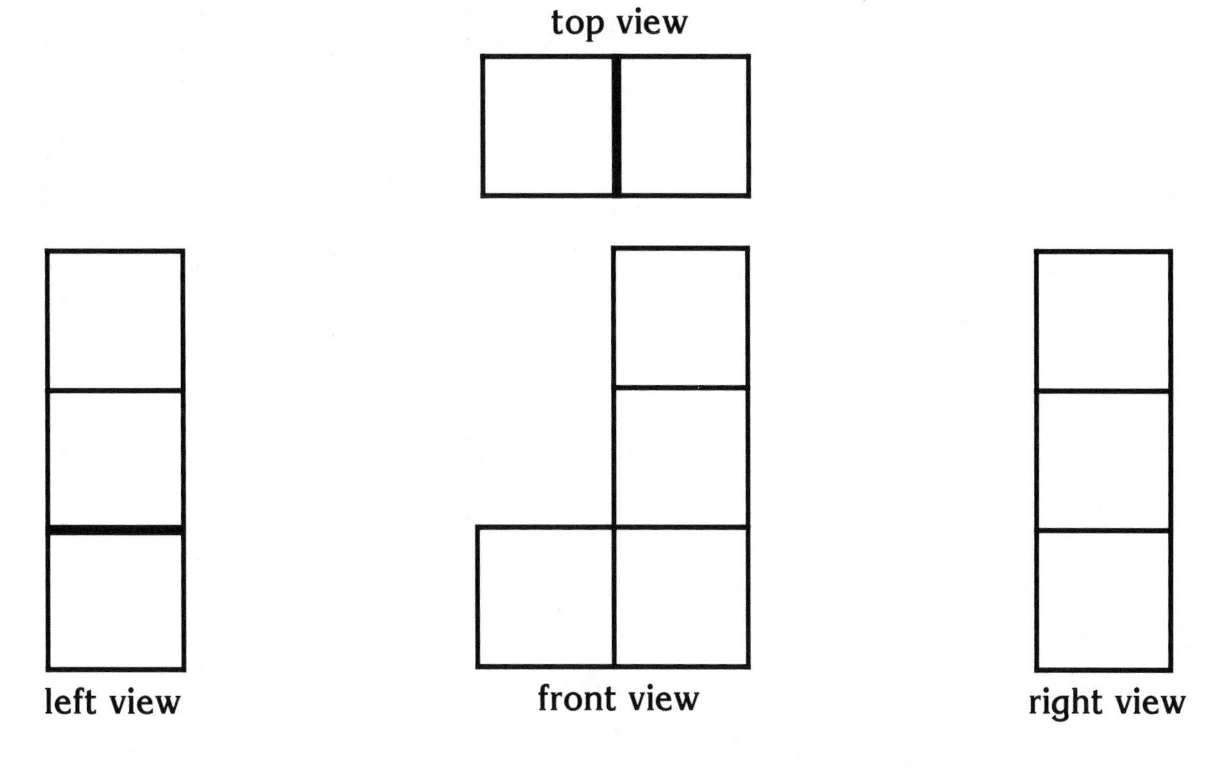

Draw the back view.

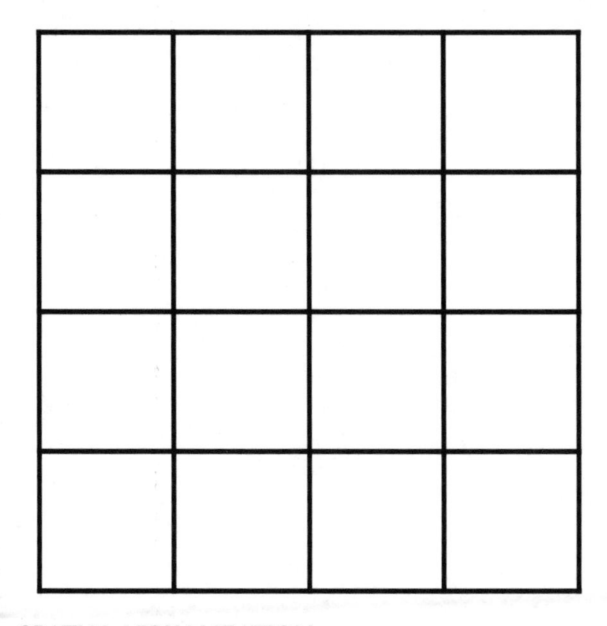

1. What is the volume?

2. What is the least perimeter and plan?

   What is the greatest perimeter and plan?

3. What is the total surface area?

4. How many vertices did you find?

Please construct the figure whose plans are shown. Consider the length of the edge of a block to measure one unit.

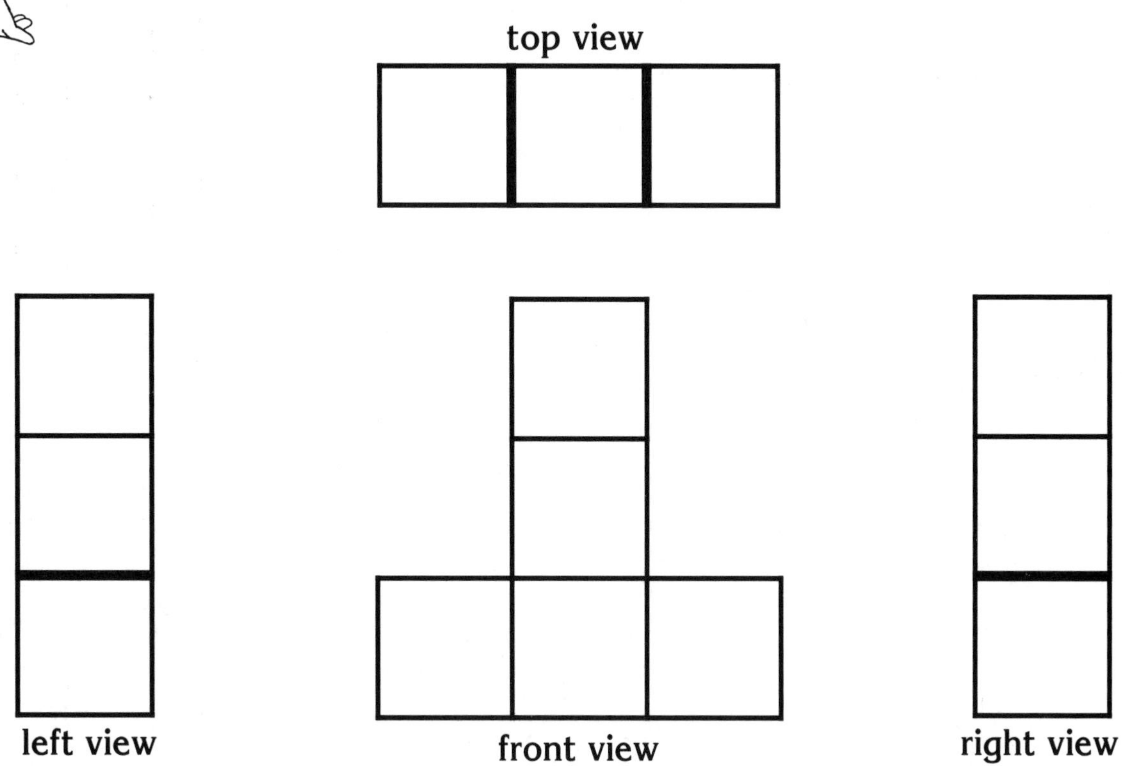

Draw the back view.

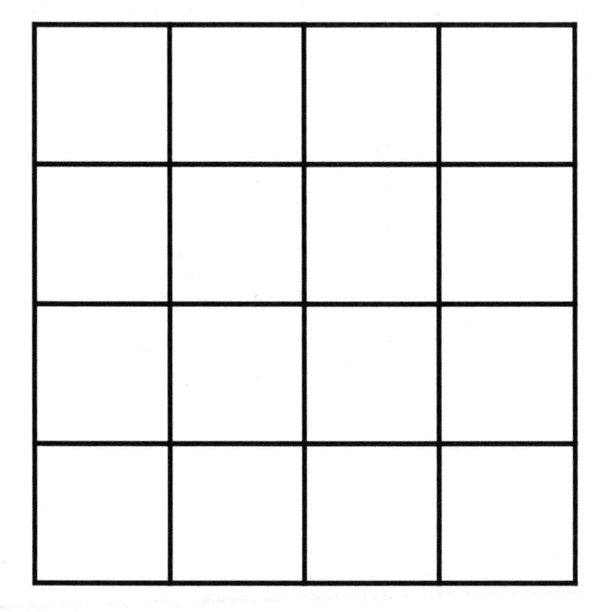

1. What is the volume?

2. What is the least perimeter and plan?

   What is the greatest perimeter and plan?

3. What is the total surface area?

4. How many vertices did you find?

Please construct the figure whose plans are shown. Consider the length of the edge of a block to measure one unit.

top view

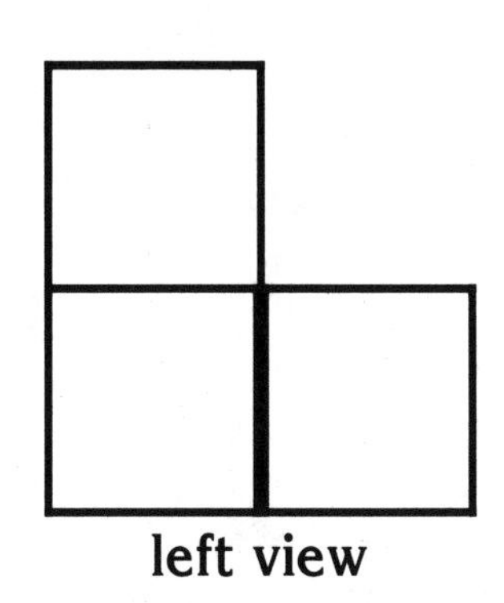

left view

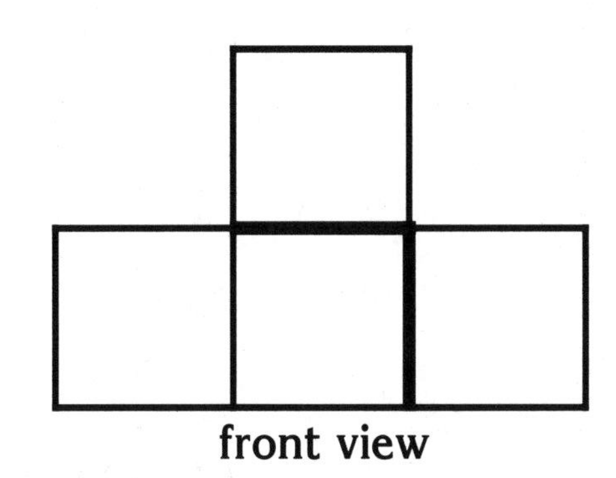

front view

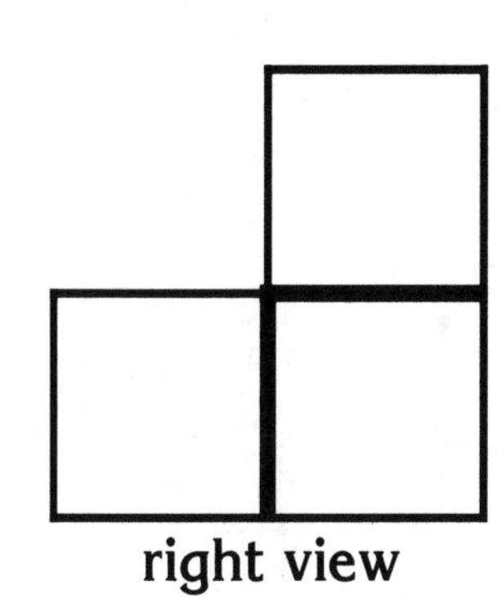

right view

Draw the back view.

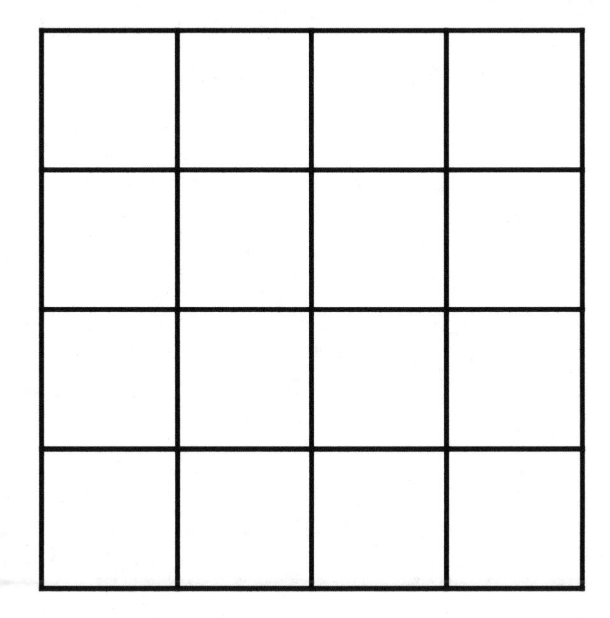

1. What is the volume?

2. What is the least perimeter and plan?

   What is the greatest perimeter and plan?

3. What is the total surface area?

4. How many vertices did you find?

Please construct the figure whose plans are shown. Consider the length of the edge of a block to measure one unit.

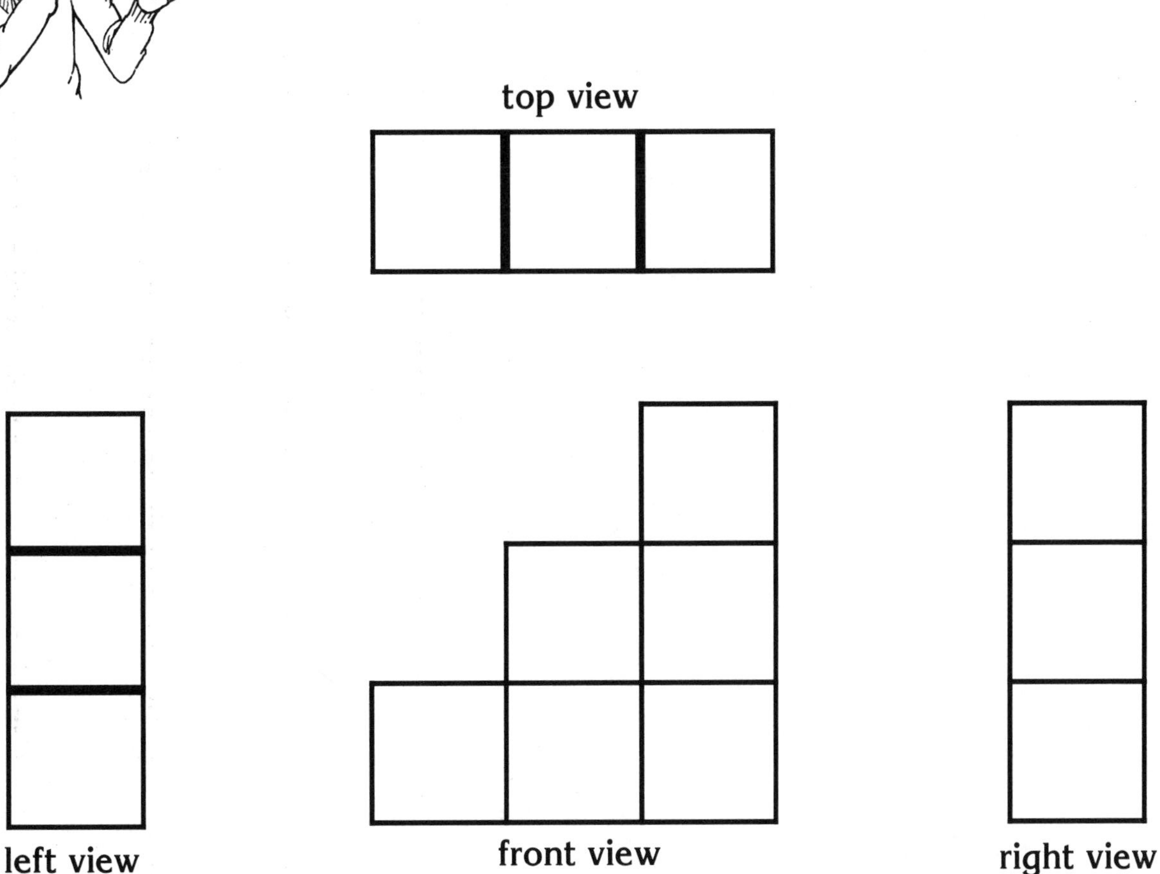

Draw the back view.

1. What is the volume?

2. What is the least perimeter and plan?

   What is the greatest perimeter and plan?

3. What is the total surface area?

4. How many vertices did you find?

Please construct the figure whose plans are shown. Consider the length of the edge of a block to measure one unit.

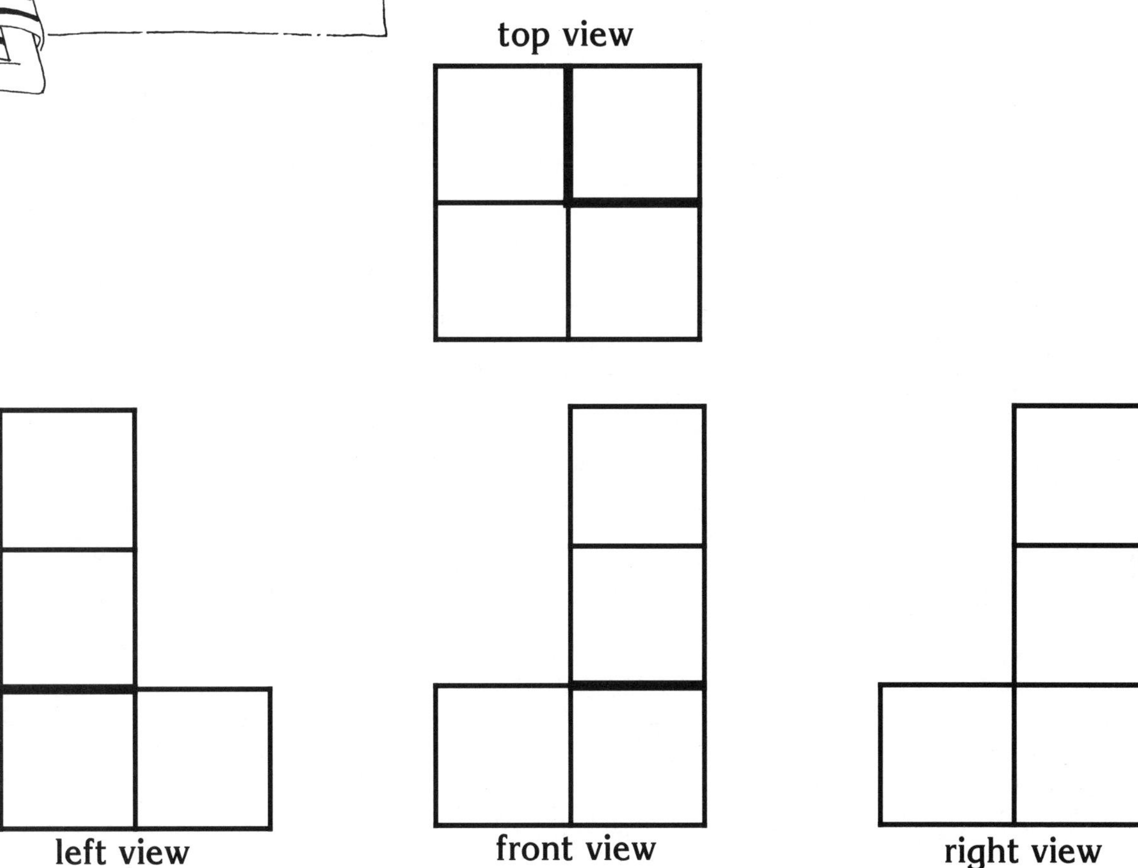

Draw the back view.

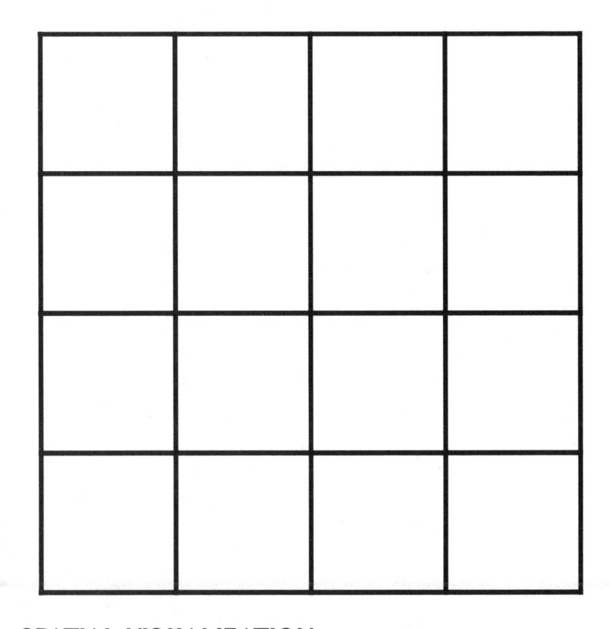

1. What is the volume?

2. What is the least perimeter and plan?

   What is the greatest perimeter and plan?

3. What is the total surface area?

4. How many vertices did you find?

Spatial Visualization

Part Two: Interpreting Isometric Drawings

In these activities students build physical models by referring to isometric drawings. They then draw their top-, front-, left-, and right-view plans. This is the inverse of the process used in *Part One*. They will continue to find volumes, surface areas, and perimeters.

The best position from which to view the model is from where only the faces being studied are seen rather than a position that provides a perspective view. This removes all of the extraneous lines from sight and reduces the possibility of confusion. Blocks which are at different distances from the viewer can be readily noted. This facilitates the drawing of the bold black lines used to indicate such differing distances.

Below are the answers to each of the activities.

top view | left view | front view | right view

Thinkcard 6
1. 6 cubic units
2. 26 square units
3. 12 units
4. 10 units

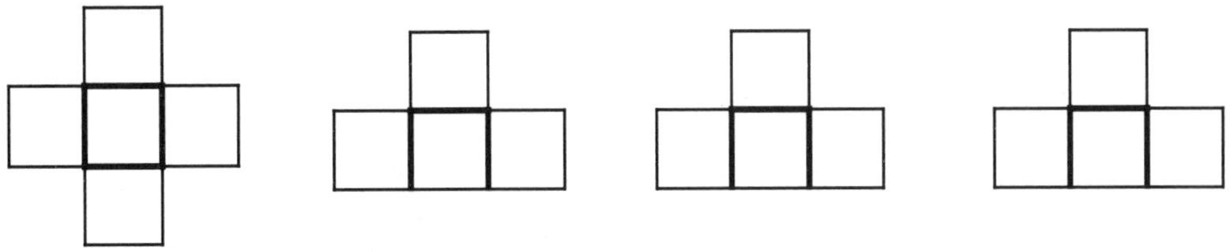

Thinkcard 7
1. 6 cubic units
2. 26 square units
3. 10 units
4. 12 units

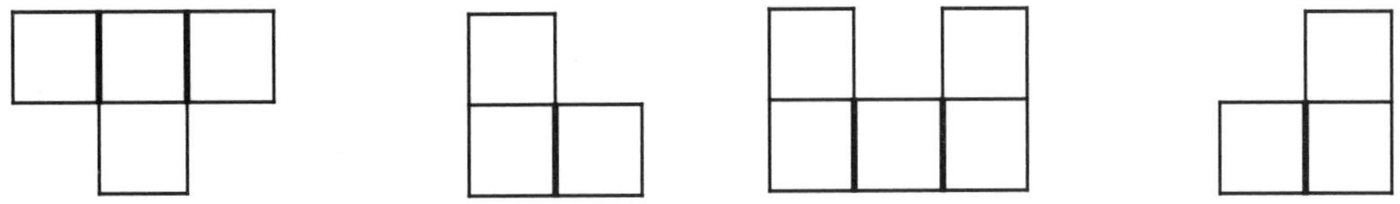

Thinkcard 8
1. 7 cubic units
2. 24 square units
3. 8 units
4. 8 units

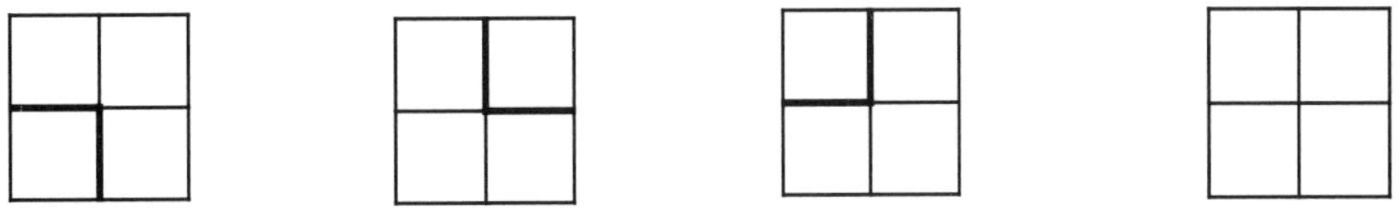

Thinkcard 9
1. 7 cubic units
2. 28 square units
3. 12 units
4. 12 units

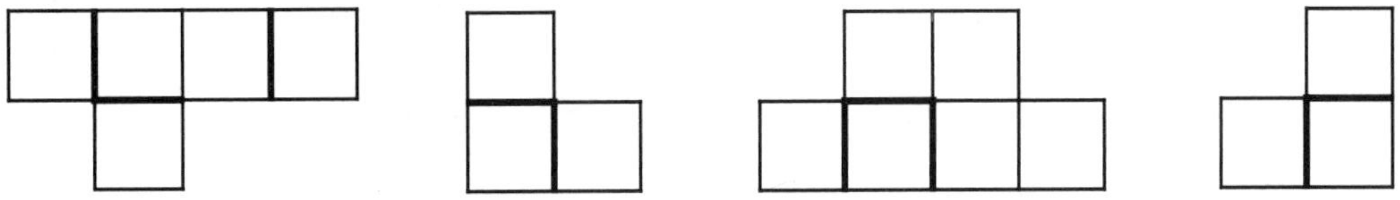

Thinkcard 10
1. 7 cubic units
2. 28 square units
3. 12 units
4. 10 units

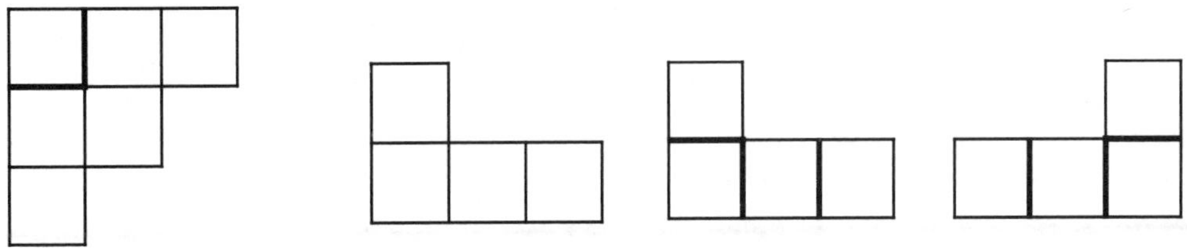

Thinkcard 6

Please construct this figure. The view from the back is the same as that of the front.

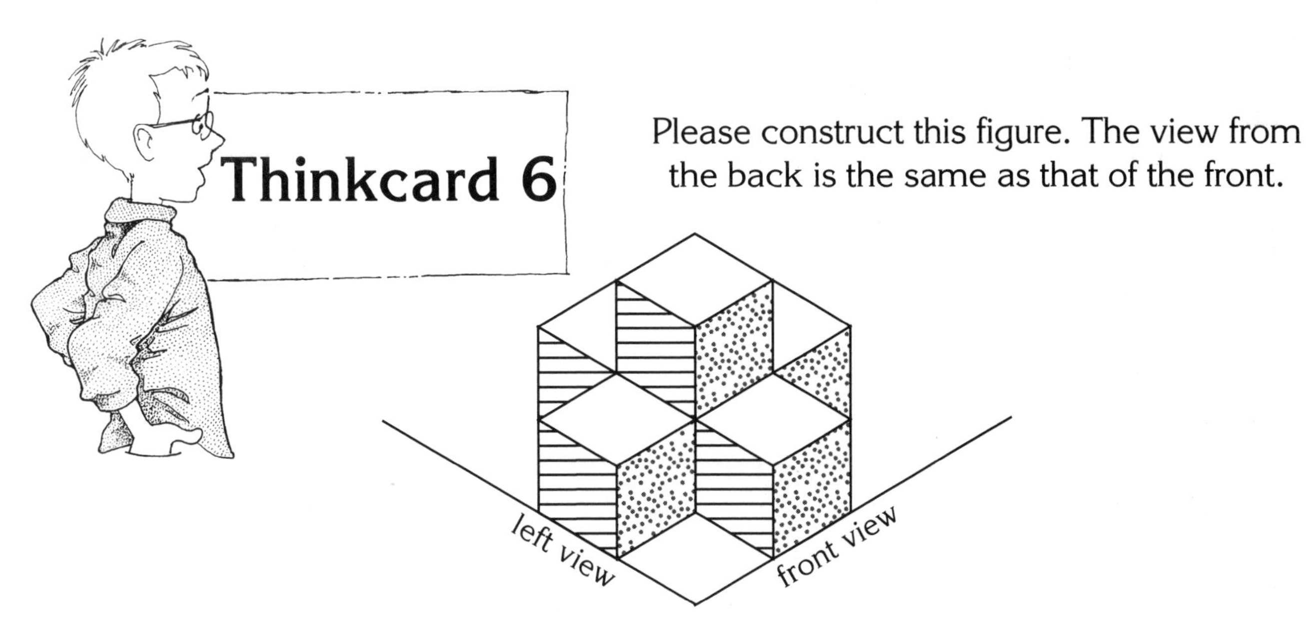

Draw the four plans below. Separate blocks which are different distances from the observer by a bold line.

top view

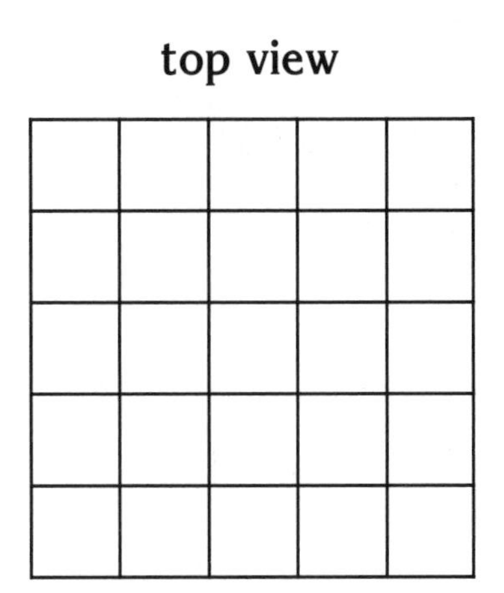

left view

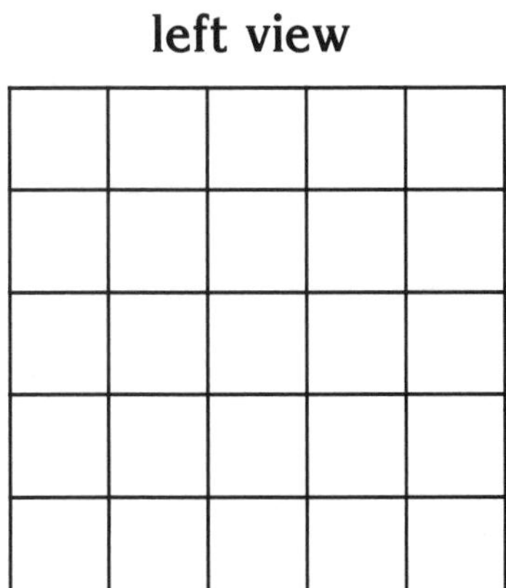
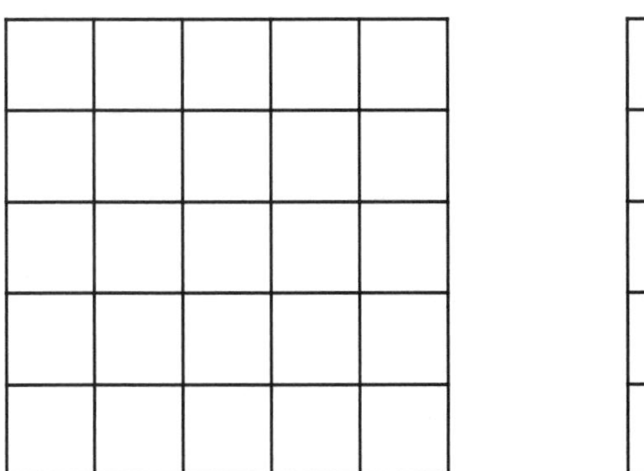

front view

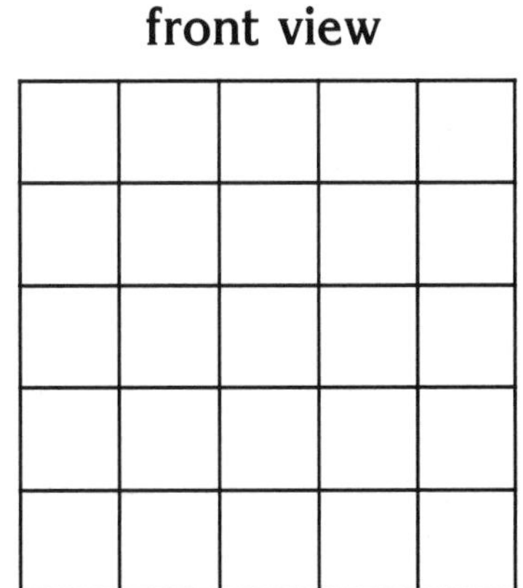
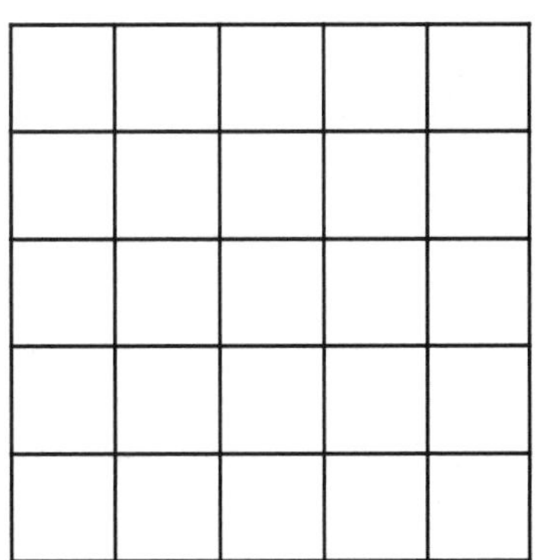

right view

1. What is the volume?
2. What is the total surface area?
3. What is the perimeter of the base?
4. What is the perimeter of the front view?

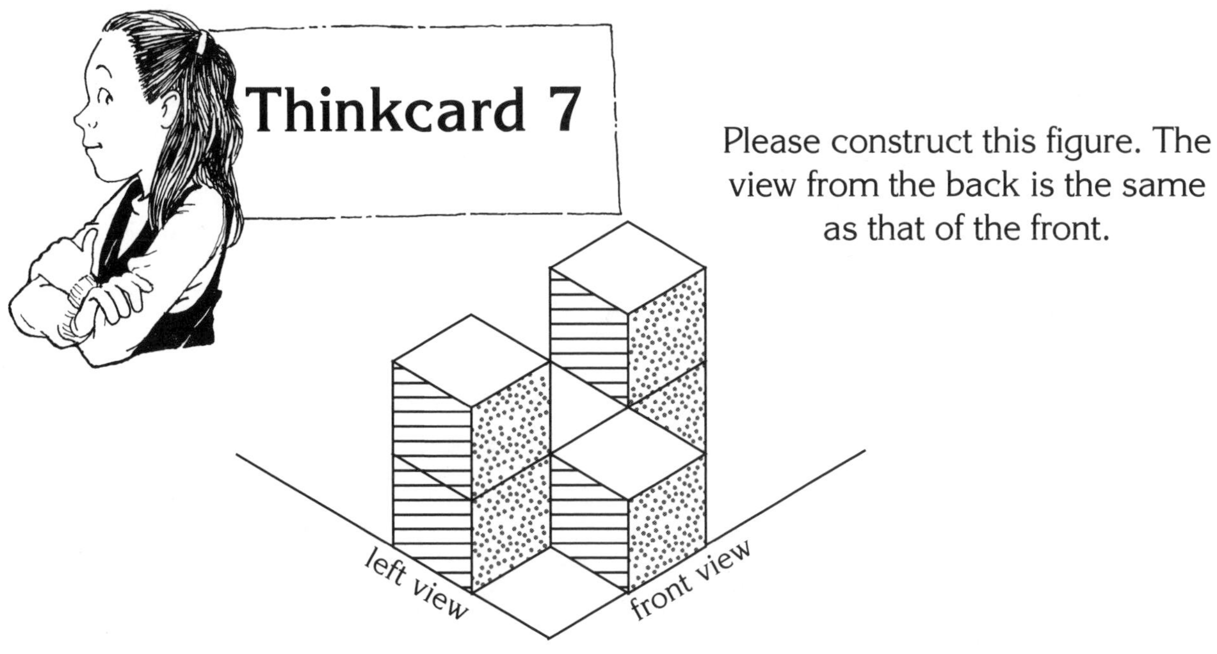

Please construct this figure. The view from the back is the same as that of the front.

Draw the four plans below. Separate blocks which are different distances from the observer by a bold line.

top view

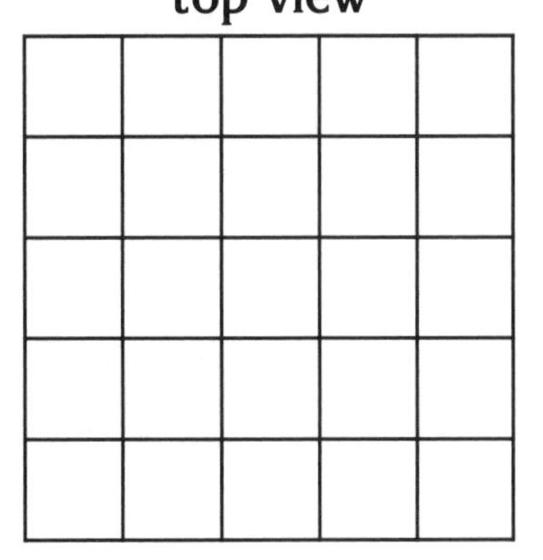

left view

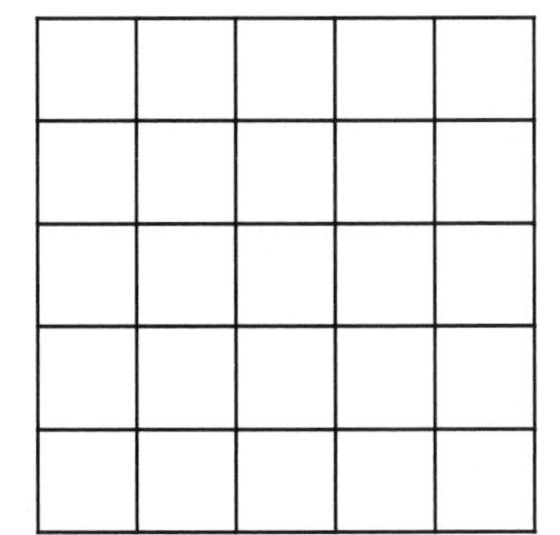

front view

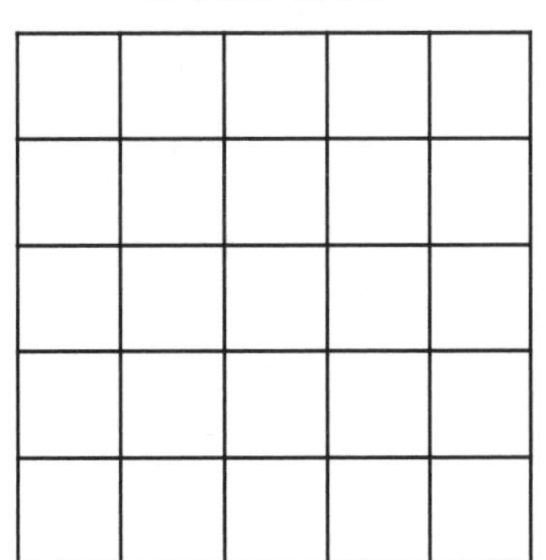

right view

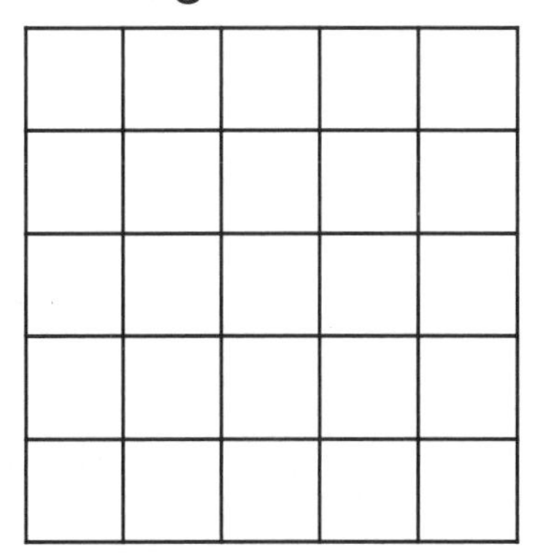

1. What is the volume?
2. What is the total surface area?
3. What is the perimeter of the base?
4. What is the perimeter of the front view?

Thinkcard 8

Please construct this figure. The view from the back is the same as that of the front.

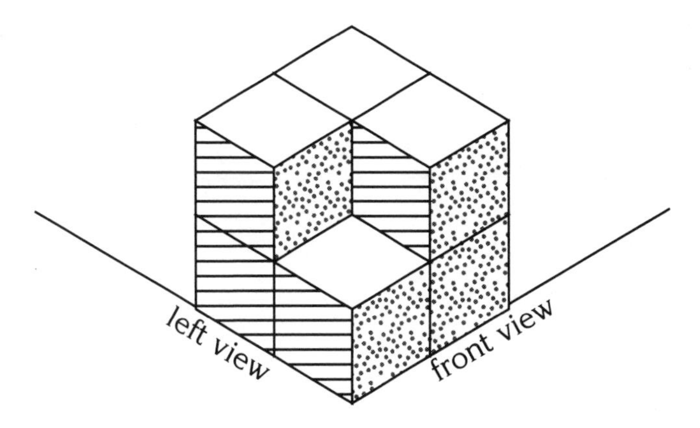

Draw the four plans below. Separate blocks which are different distances from the observer by a bold line.

top view

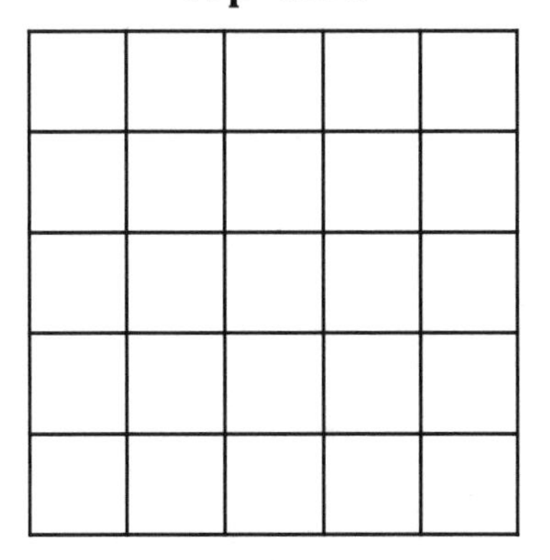

left view

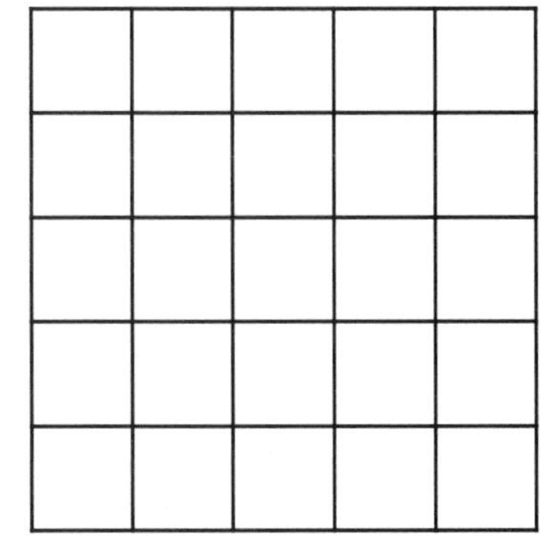

front view

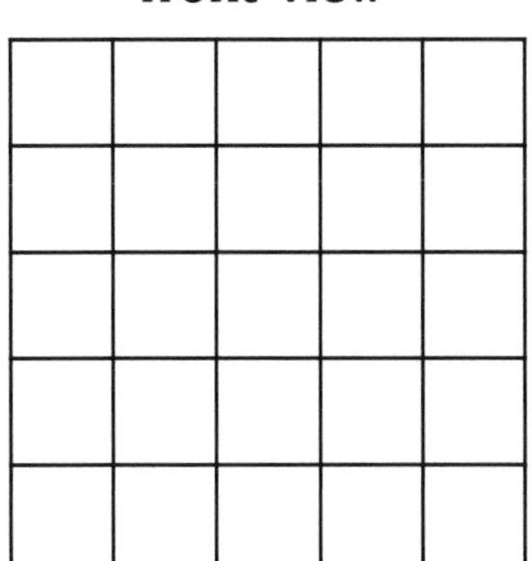

right view

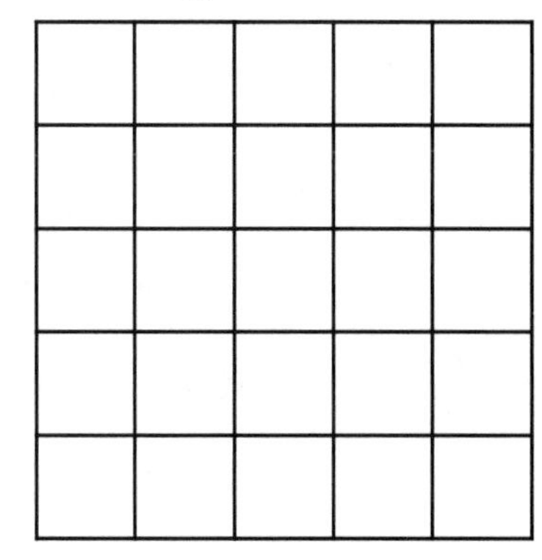

1. What is the volume?
2. What is the total surface area?
3. What is the perimeter of the base?
4. What is the perimeter of the front view?

Please construct this figure. The view from the back is the same as that of the front.

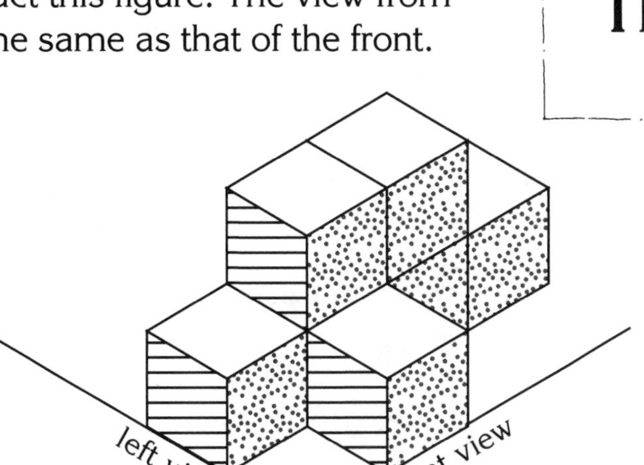

Draw the four plans below. Separate blocks which are different distances from the observer by a bold line.

top view

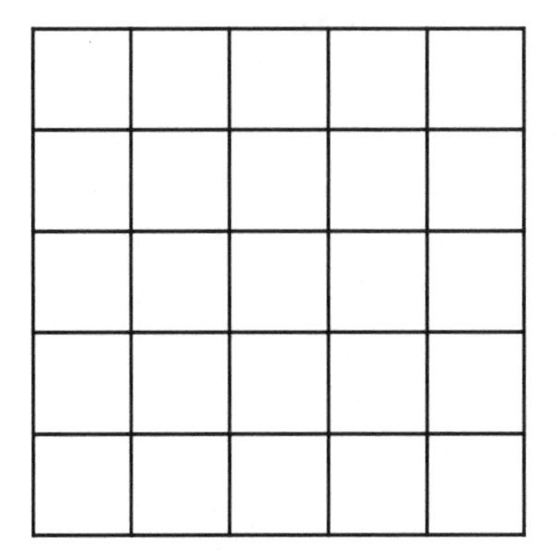

left view

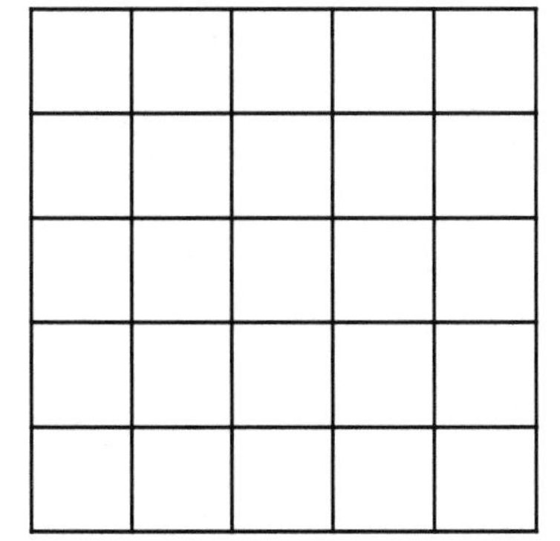

front view

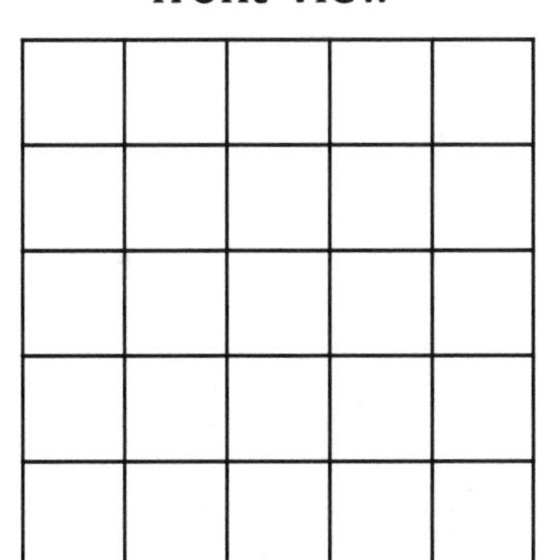

right view

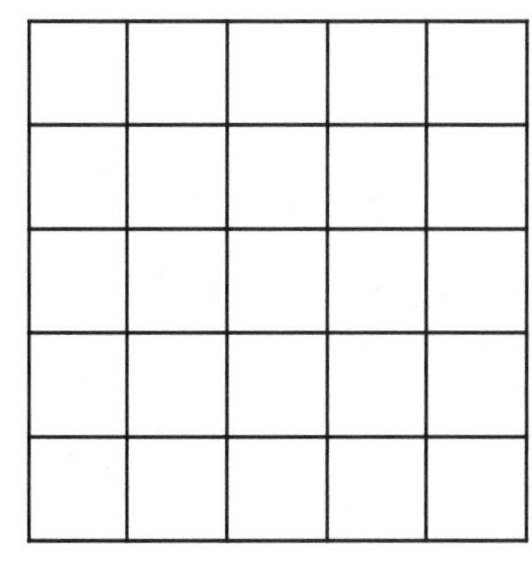

1. What is the volume?
2. What is the total surface area?
3. What is the perimeter of the base?
4. What is the perimeter of the front view?

Please construct this figure. The view from the back is the same as that of the front.

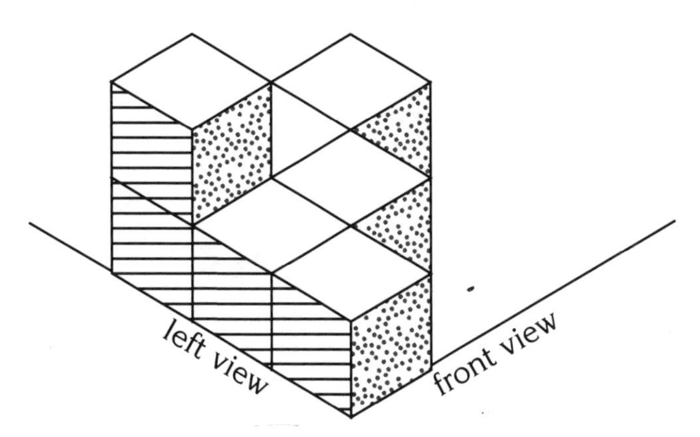

Draw the four plans below. Separate blocks which are different distances from the observer by a bold line.

top view

left view

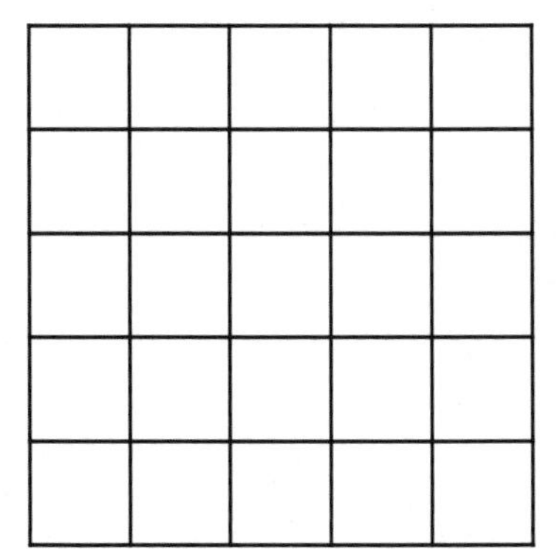

front view

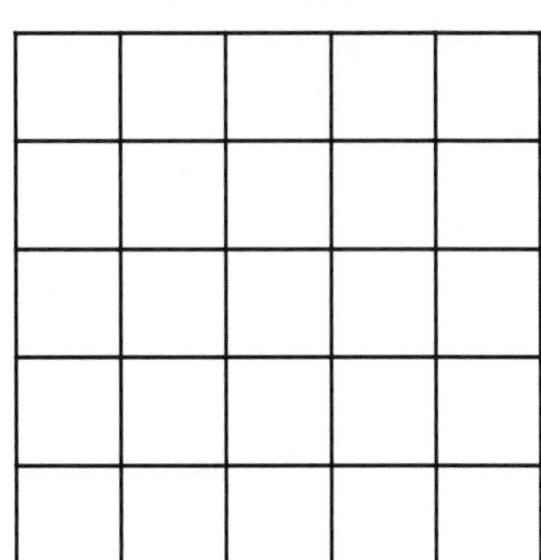

right view

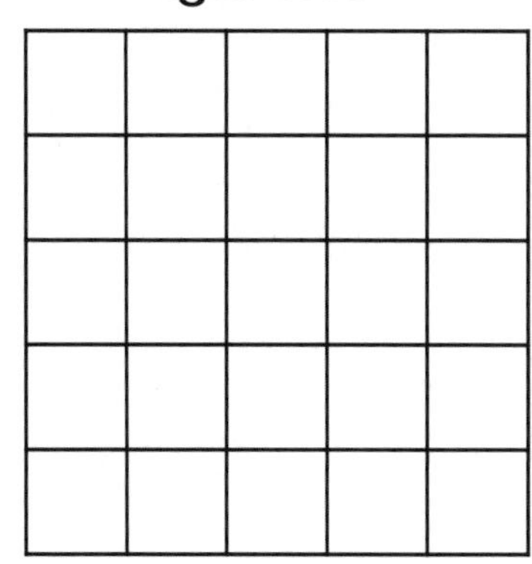

1. What is the volume?
2. What is the total surface area?
3. What is the perimeter of the base?
4. What is the perimeter of the front view?

Spatial Visualization

Part Three: Making and Interpreting Isometric Drawings

Isometric and "exploded" view drawings are used extensively in many professions and industries. Facility in creating and interpreting such drawings is a valuable life skill.

It is advisable to carefully guide students through their orientation to making isometric drawings. One method is for the teacher to demonstrate each step using the overhead while students are simultaneously duplicating each step.

Using AIMS Spatial Visualization Tiles

The AIMS Spatial Visualization Tiles are excellent manipulatives for acquainting students with the components used in isometric drawings (rhombi, or diamonds, and equilateral triangles) and how they are arranged to produce a three-dimensional effect. They open wide the door to the exploration creating new designs from which models can be constructed. Since they are transparent, they can be used on the overhead projector to model the process of making isometric drawings.

If these tiles are not available, a grid of these components is included in this section. Duplicate the grid using three colors of paper and cut apart to provide the diamond and triangular shapes.

Provide the simplest possible introduction to the class by showing how one cube is constructed. They should learn that the isometric view of a cube is actually a regular hexagon with each diamond face a different color to give it perspective. Have them study the orientation of each diamond. The major diagonal of the top is always horizontal, the one of the left face always slants to the left as it rises, and the one on the right face always slants to the right as it rises.

With either manipulatives, tiles or paper, stress that the same color is used for all tops, another for left faces, and a third for right faces.

As soon as possible, free the students to explore making isometric representations with the tiles or colored paper components. They can learn much on their own. Next, introduce them to making isometric drawings which will permit them to make a record of their designs.

The problem-solving activities on *Thinkcards* 11, 12, 13, and 14 ask students to use the manipulative to fill the outline in such a way that is an isometric or perspective representation of a model. These activities may be interspersed with the others in this *Part Three* as deemed most appropriate. They add an interesting variation to the processes used in the other activities.

Introducing Isometric Drawings

The recommended starting point for introducing students to making isometric drawings is to tutor students through two views of a single cube and to have them draw the cube from several additional perspectives.

A first step is to draw two diamonds as shown. Whether these become the tops or bottoms of cubes depends on how vertical lines are drawn from the vertices. If lines are drawn downward, the face becomes a top; if they are drawn upward, it becomes a bottom. After this is understood, ask students to draw the cube from as many perspectives as possible.

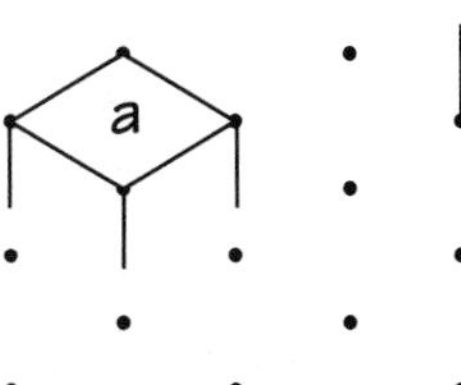

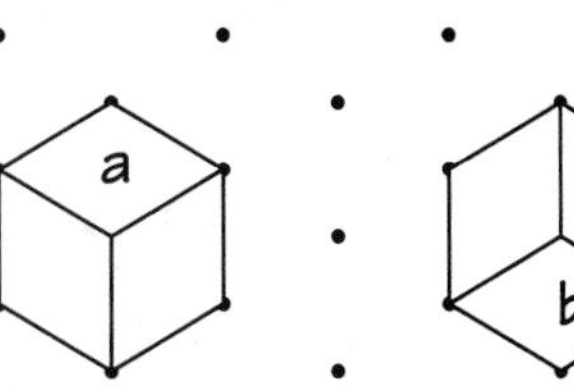

The next step is to ask students to snap two cubes together and draw the resulting figure from several perspectives.

Following this introduction, students should be ready to tackle the sequence of shapes in *Thinkcards 18* and *19*. Patience and a sufficient allocation of time are particularly important as students build their ability with these exercises. Some will need additional tutoring as they work through increasingly complex shapes.

Four selected perspectives are shown for each of the shapes.

After these preliminary exercises, students should be able to launch into the more challenging designs presented in *Thinkcards 20-24.*

Answers 12-17

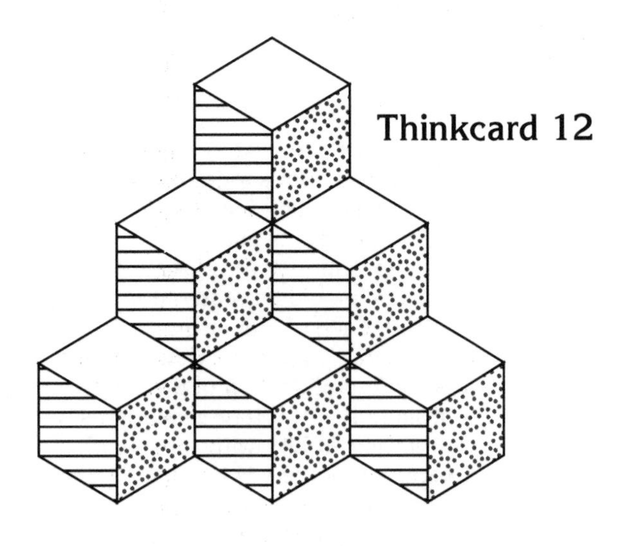

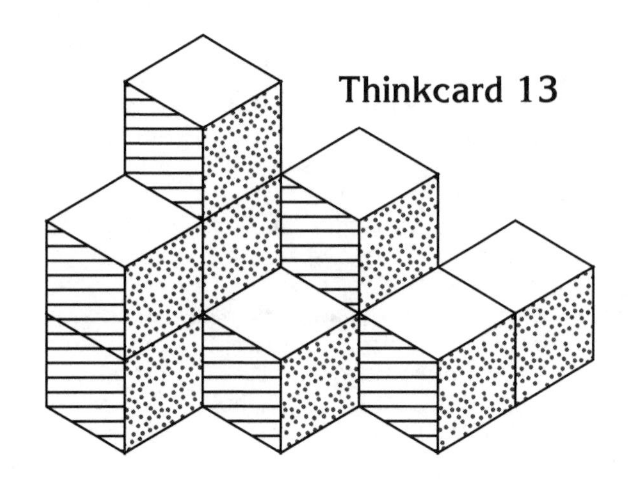

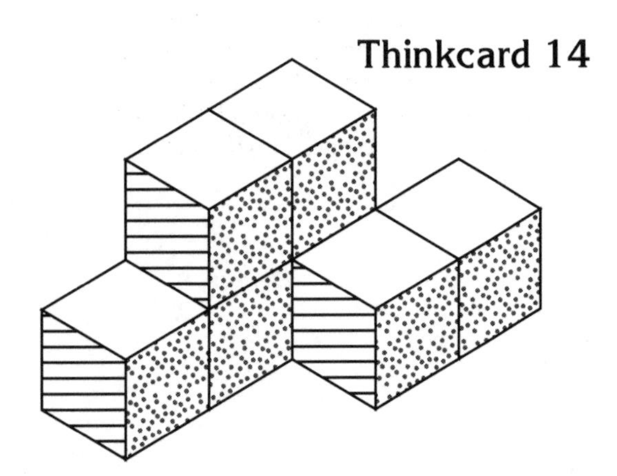

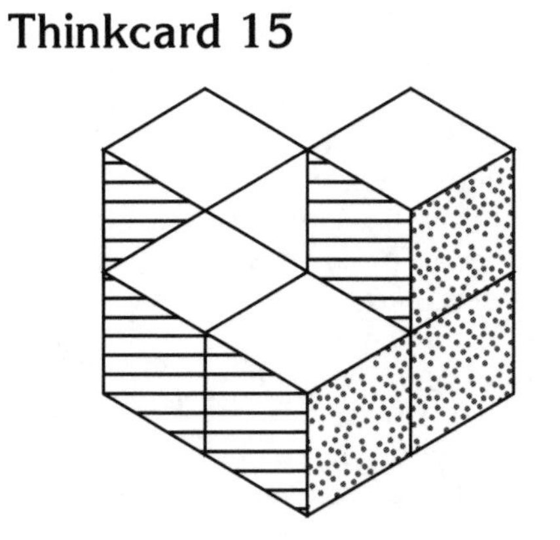

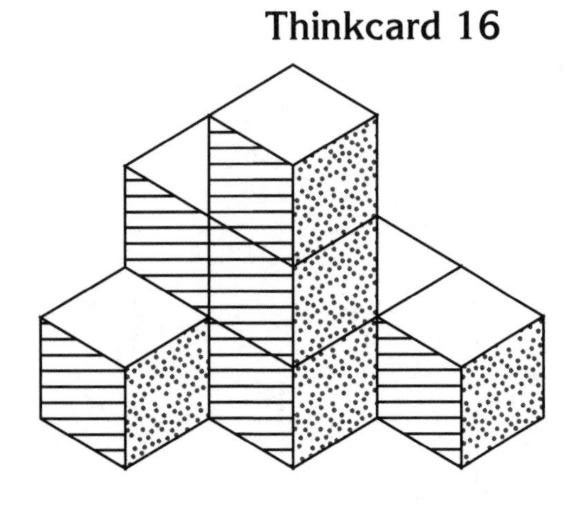

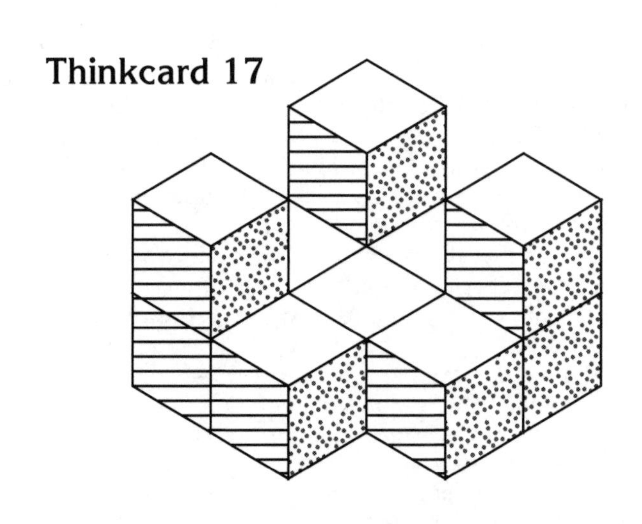

Answers 18a-19b

Thinkcard 18a

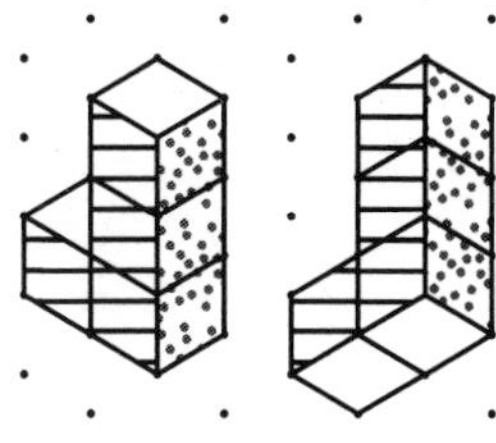

Thinkcard 18b

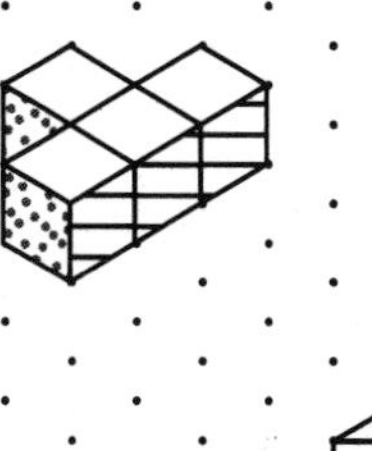

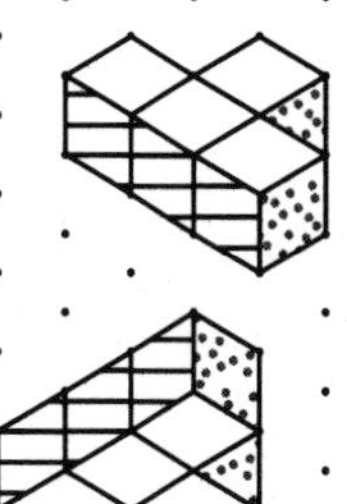

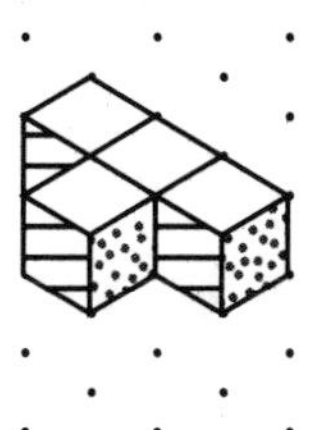

Thinkcard 19a

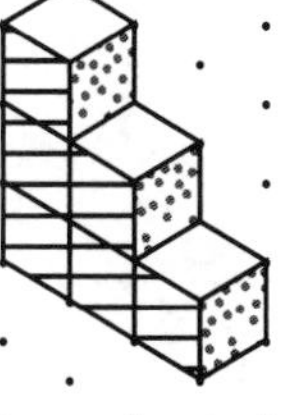

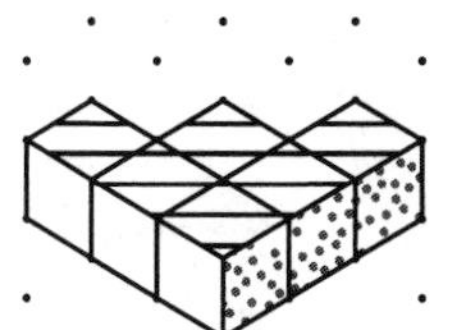

Thinkcard 19b

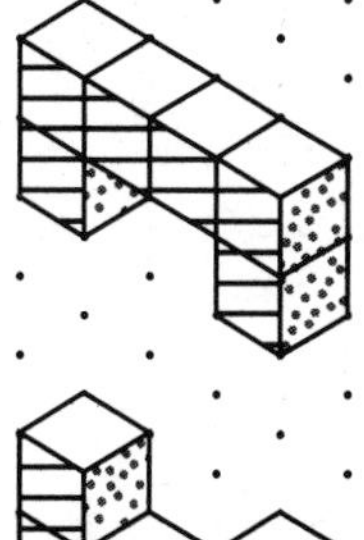

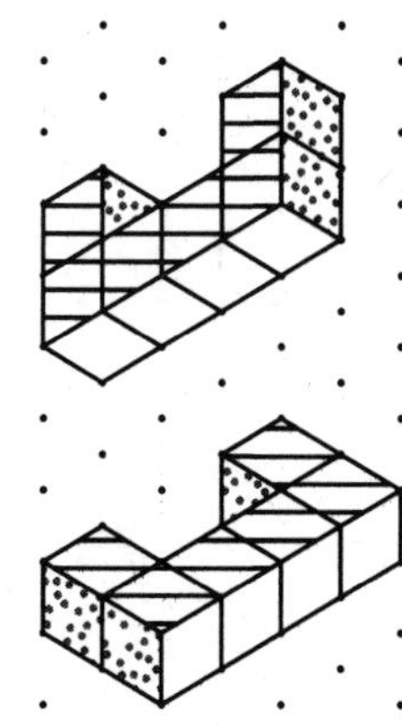

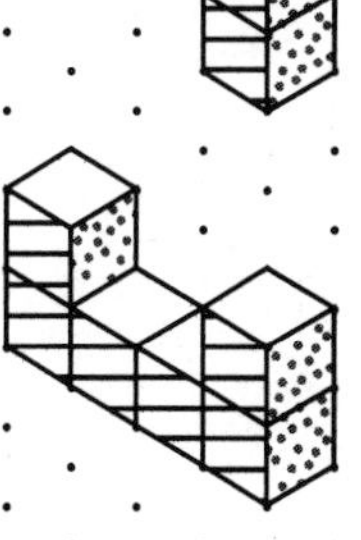

Answers for Thinkcards 20-24

Students may make the isometric drawings from any perspective they wish. Therefore, the following drawings are only examples of those possible.

Thinkcard 20

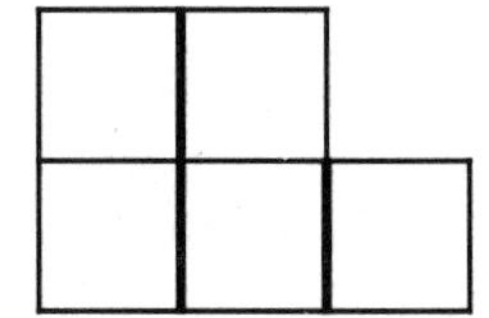

1. 7 cubic units
2. 12 units
3. 30 square units

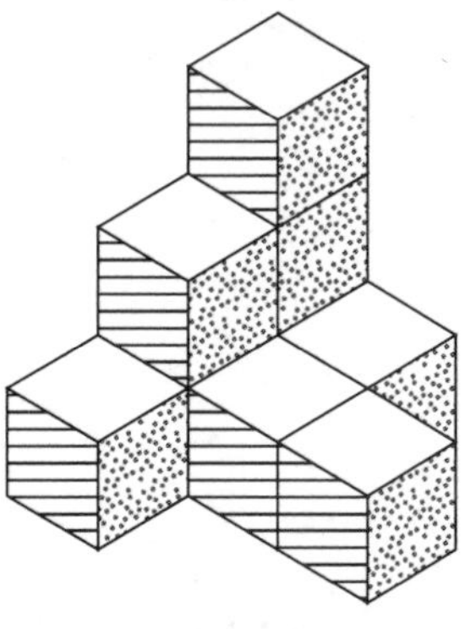

Thinkcard 21

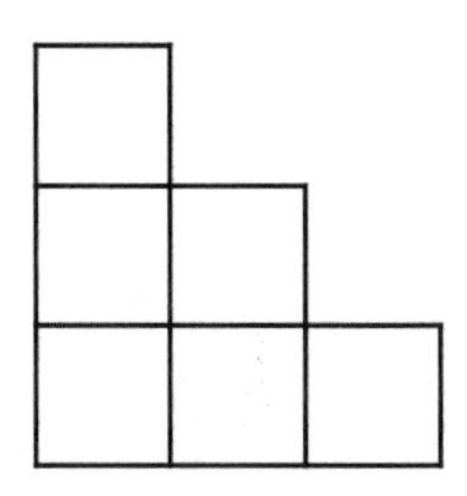

1. 9 cubic units
2. 12 units
3. 34 square units

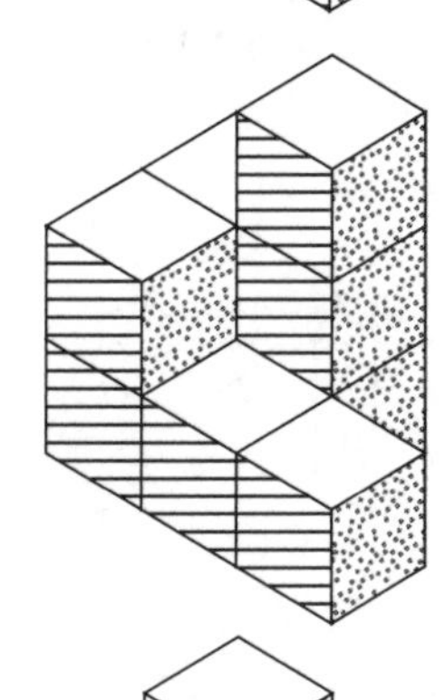

Thinkcard 22

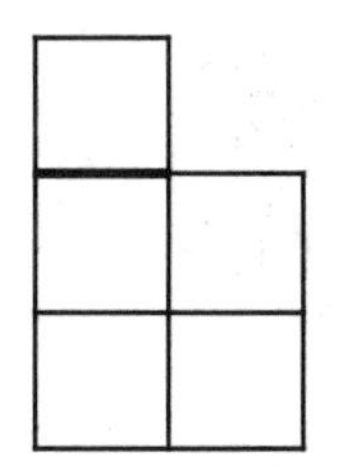

1. 9 cubic units
2. 10 units
3. 32 square units

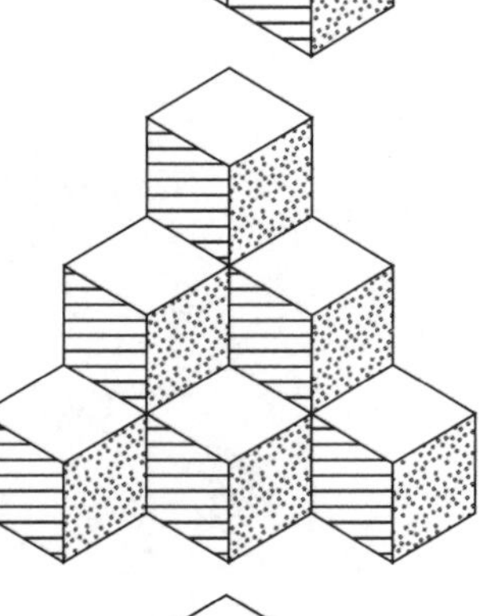

Thinkcard 23

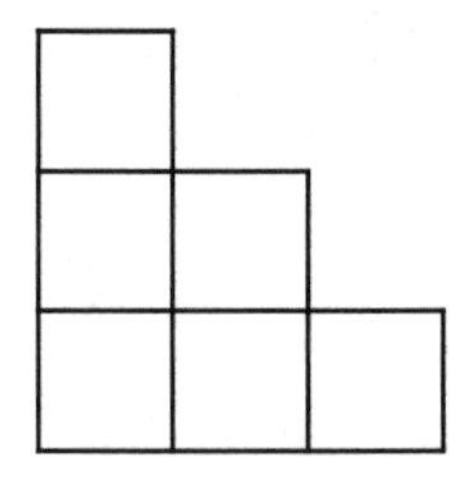

1. 10 cubic units
2. 12 units
3. 36 square units

Thinkcard 24

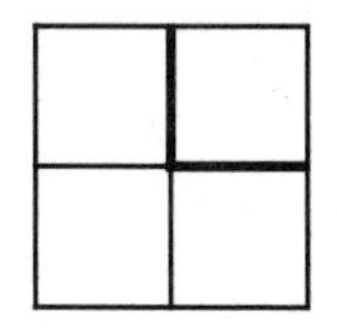

1. 6 cubic units
2. 8 units
3. 24 square units

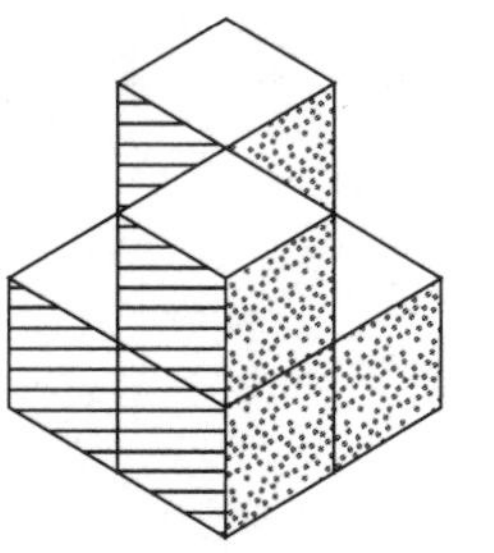

Thinkcard 11

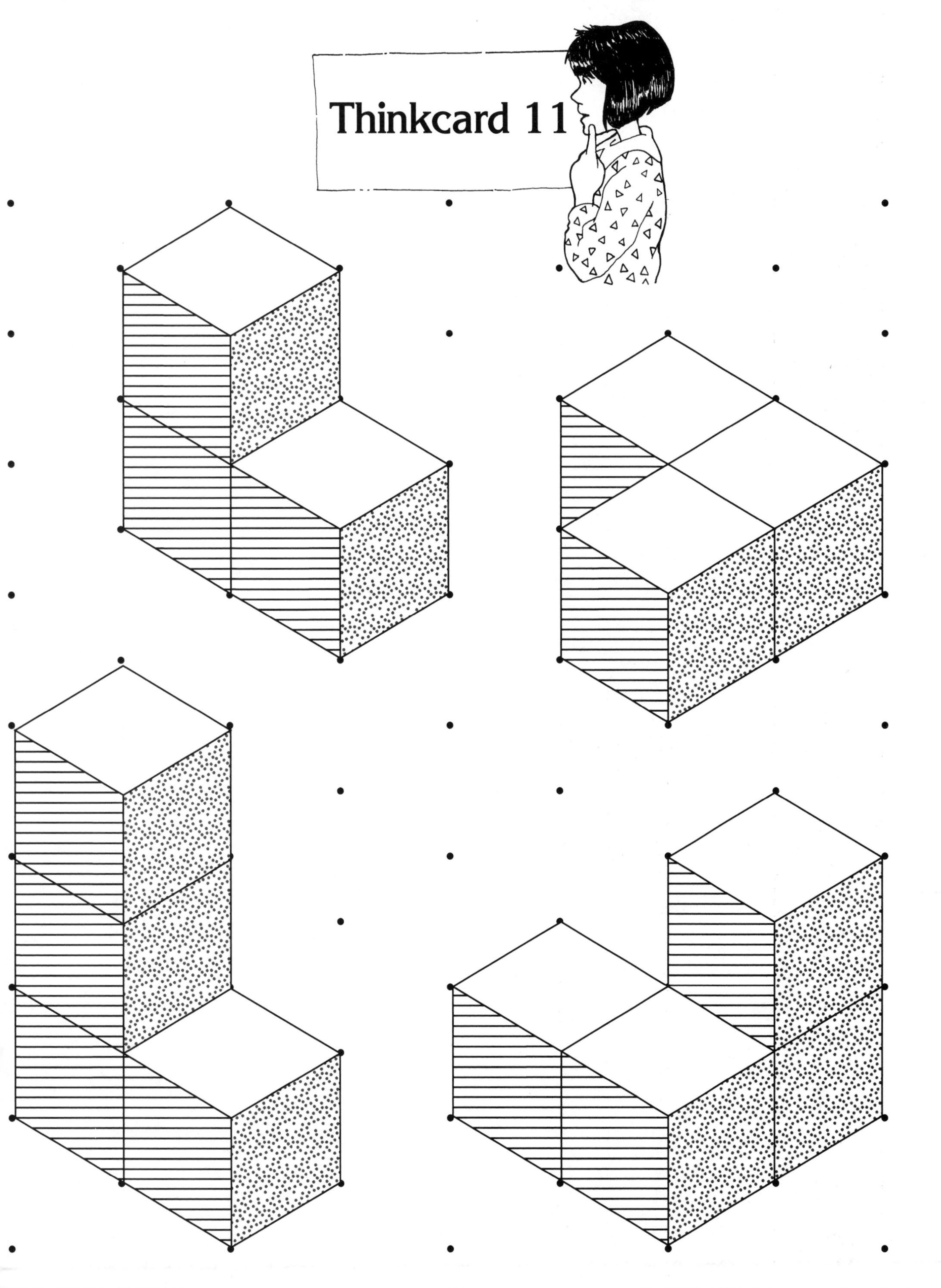

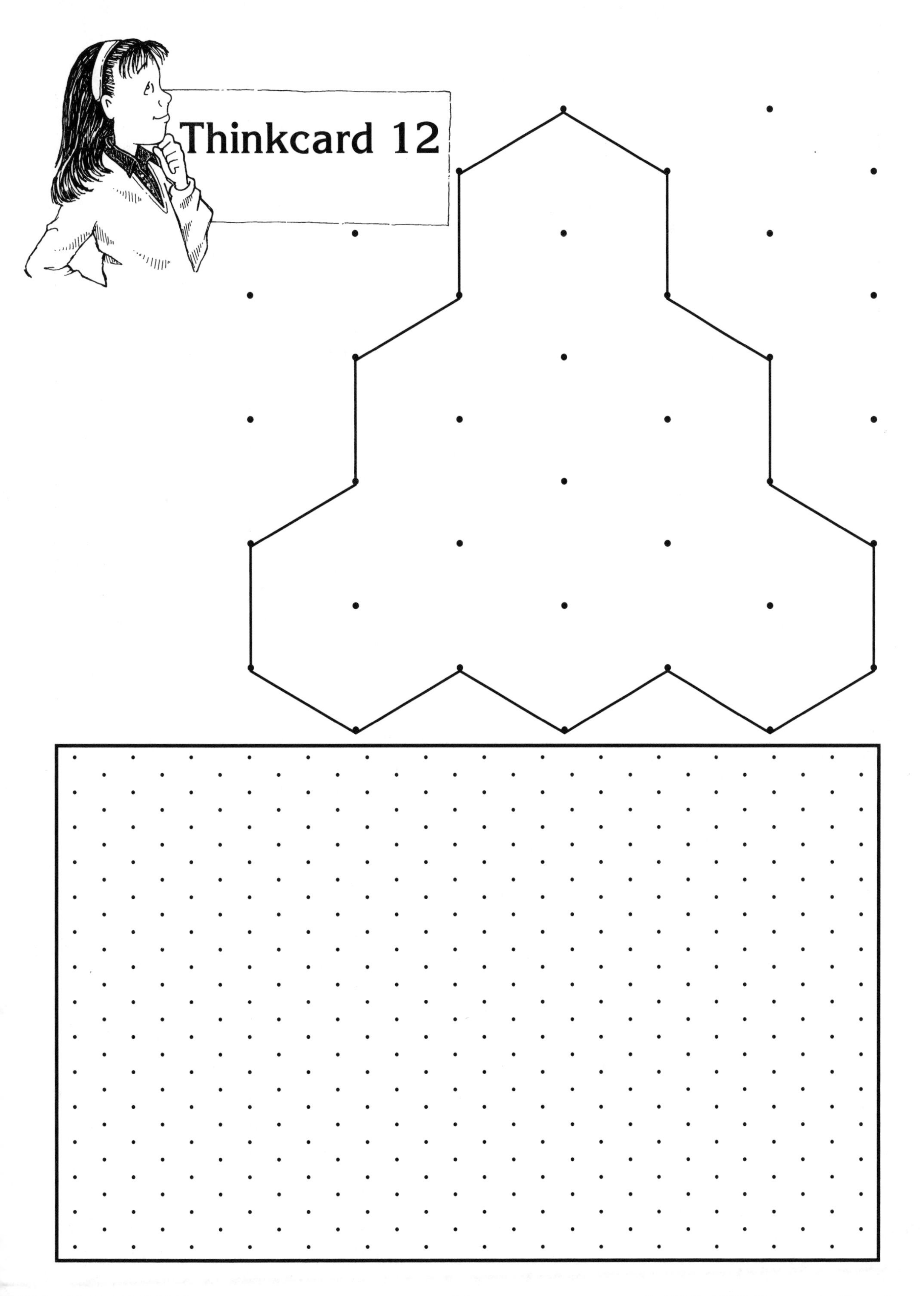
Thinkcard 12

Thinkcard 13

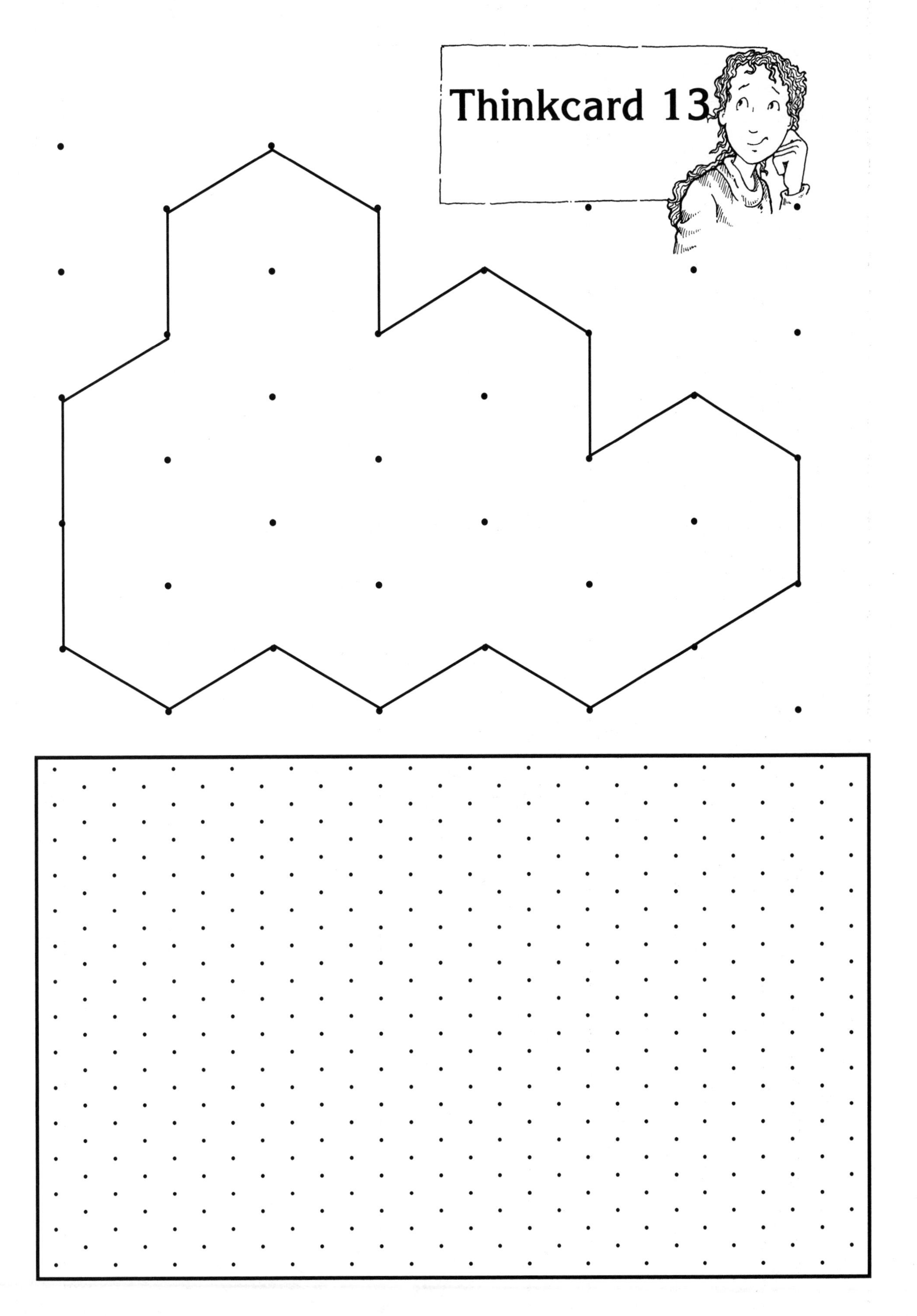

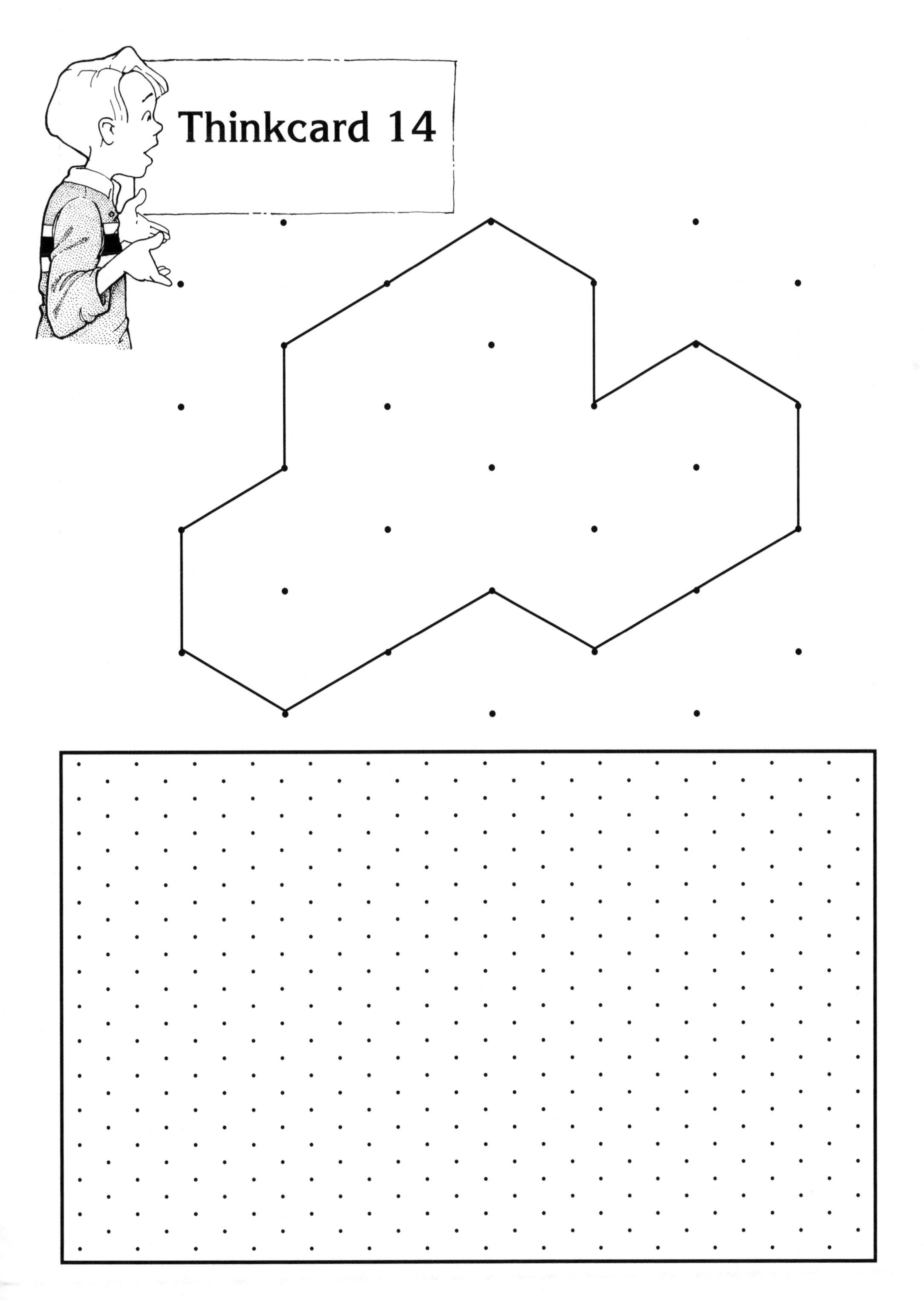
Thinkcard 14

Thinkcard 15

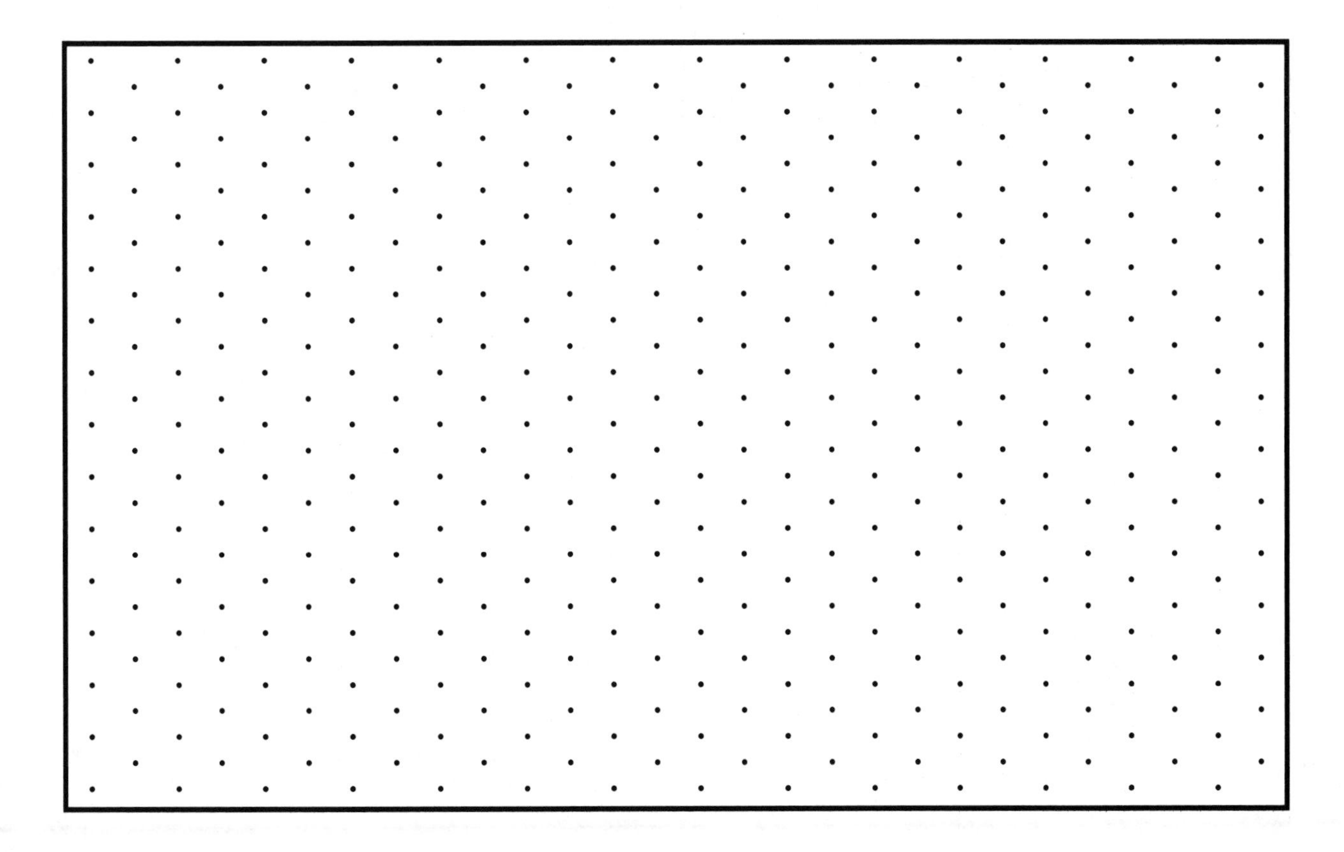

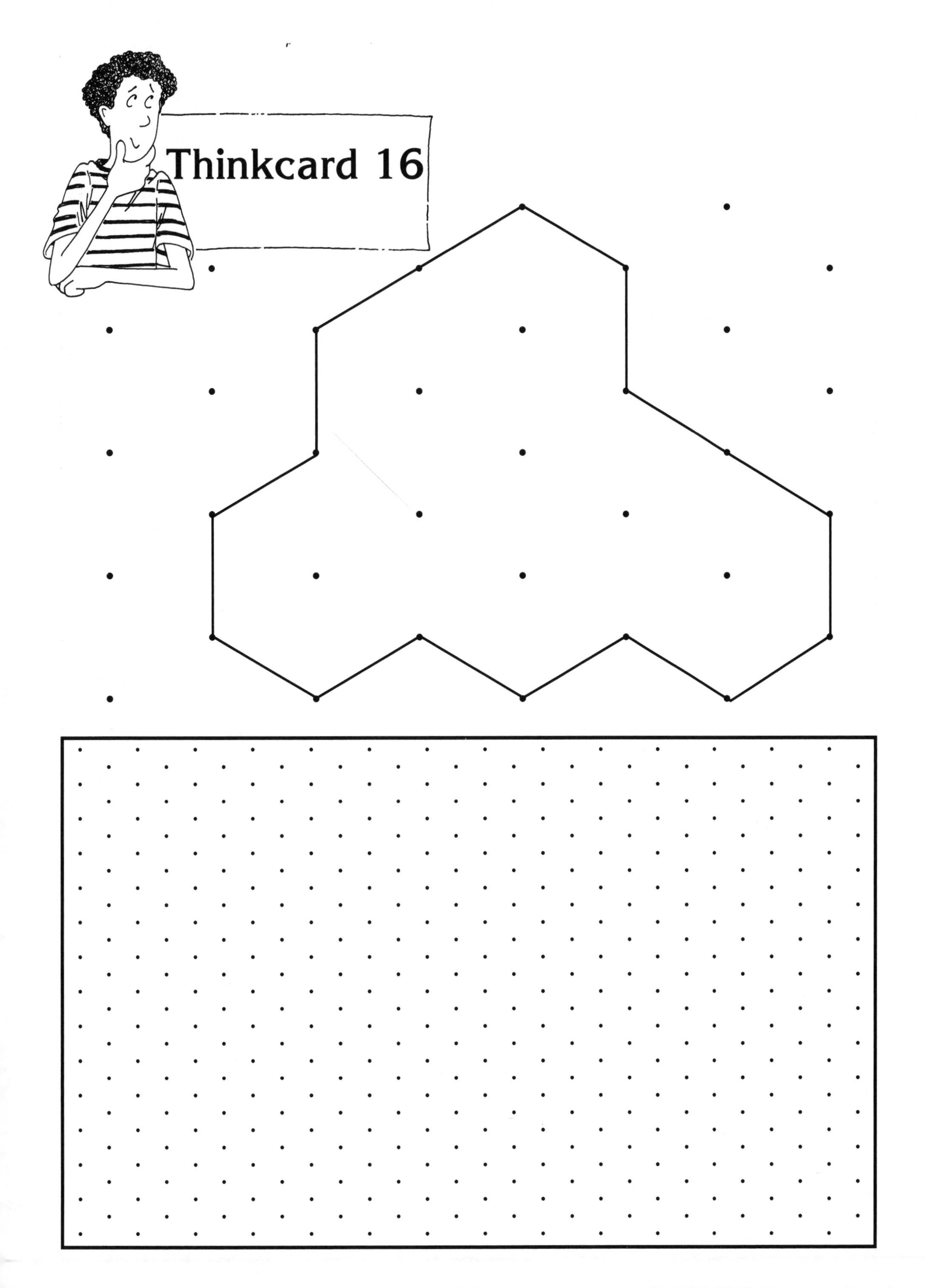
Thinkcard 16

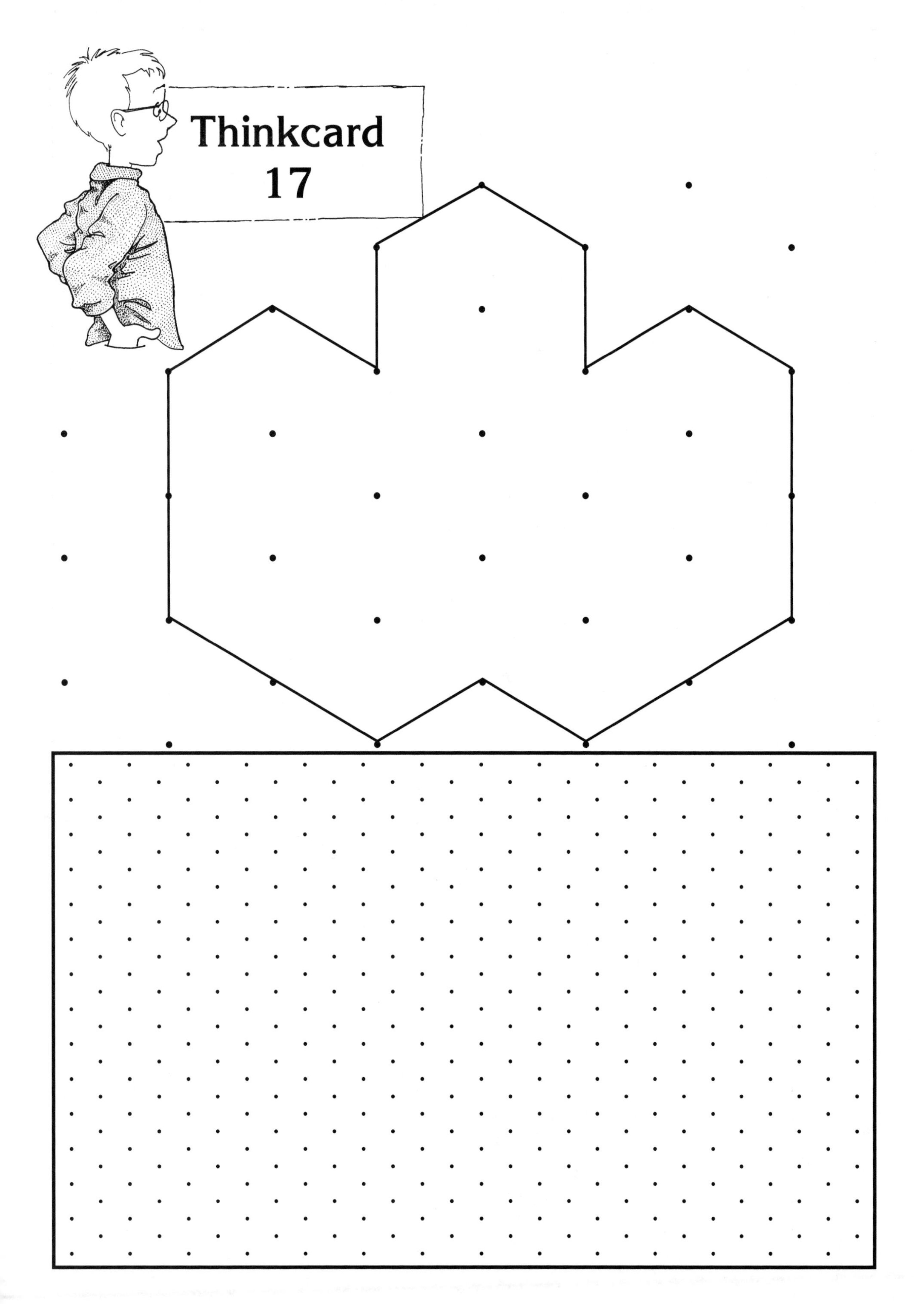
Thinkcard
17

Please construct each of these figures. Then draw four more views of each from different perspectives.

18a

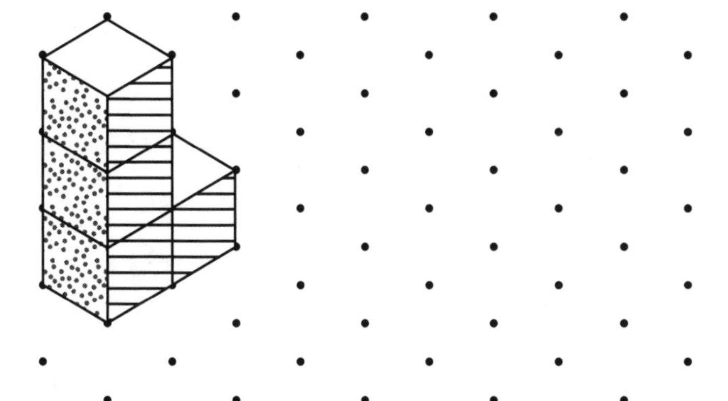

18b

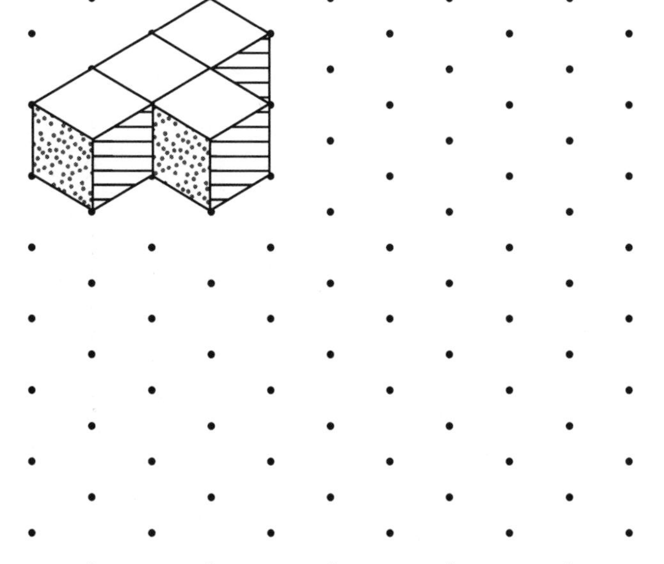

Thinkcard 19

Please construct each of these figures. Then draw four more views of each from different perspectives.

19a

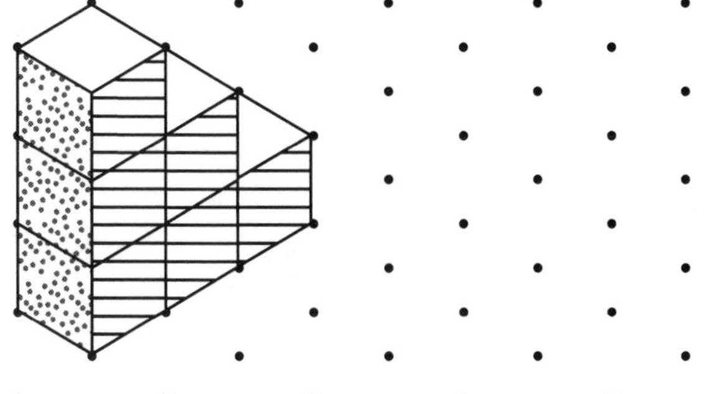

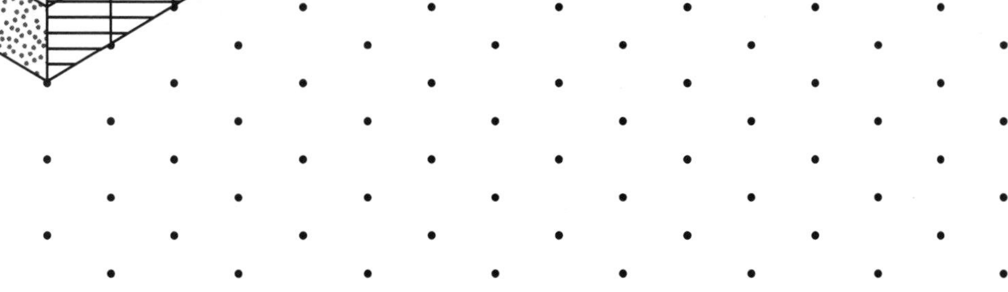

19b

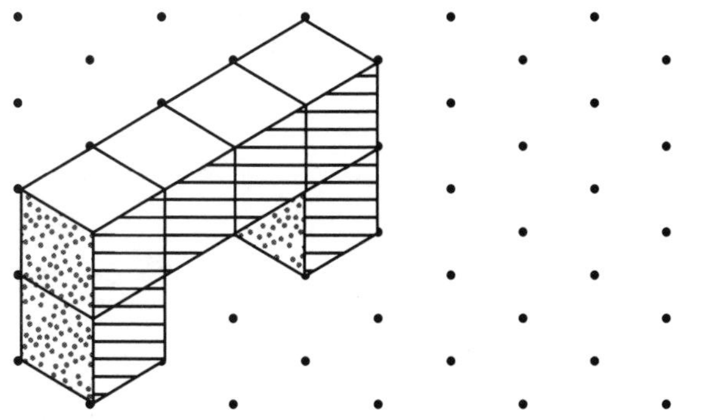

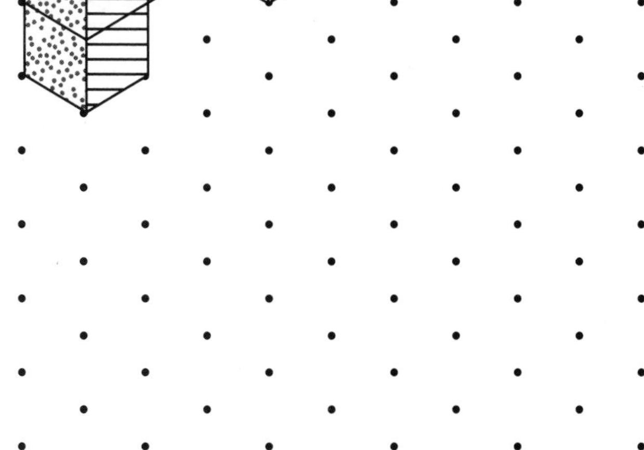

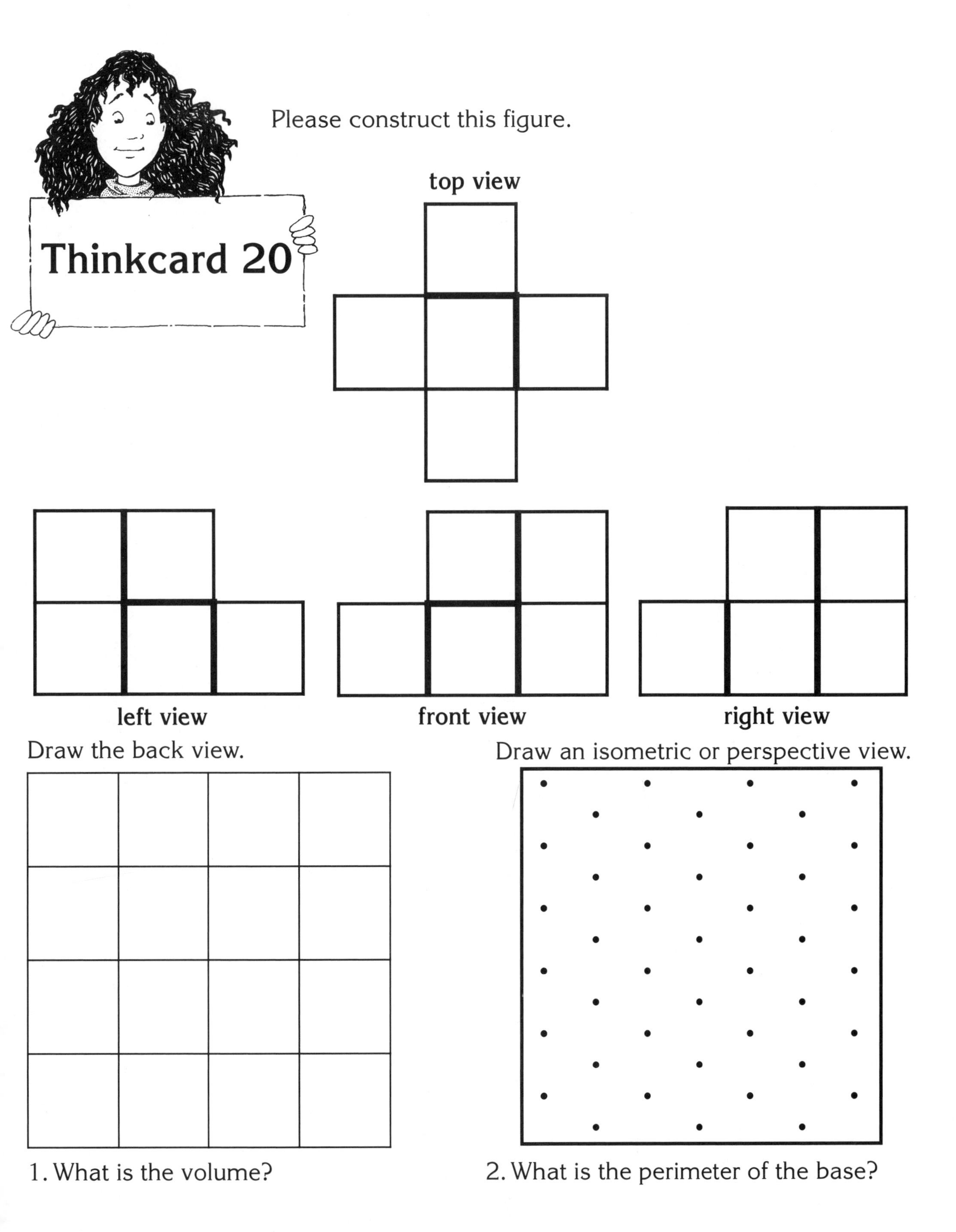

1. What is the volume?

2. What is the perimeter of the base?

3. What is the total surface area?

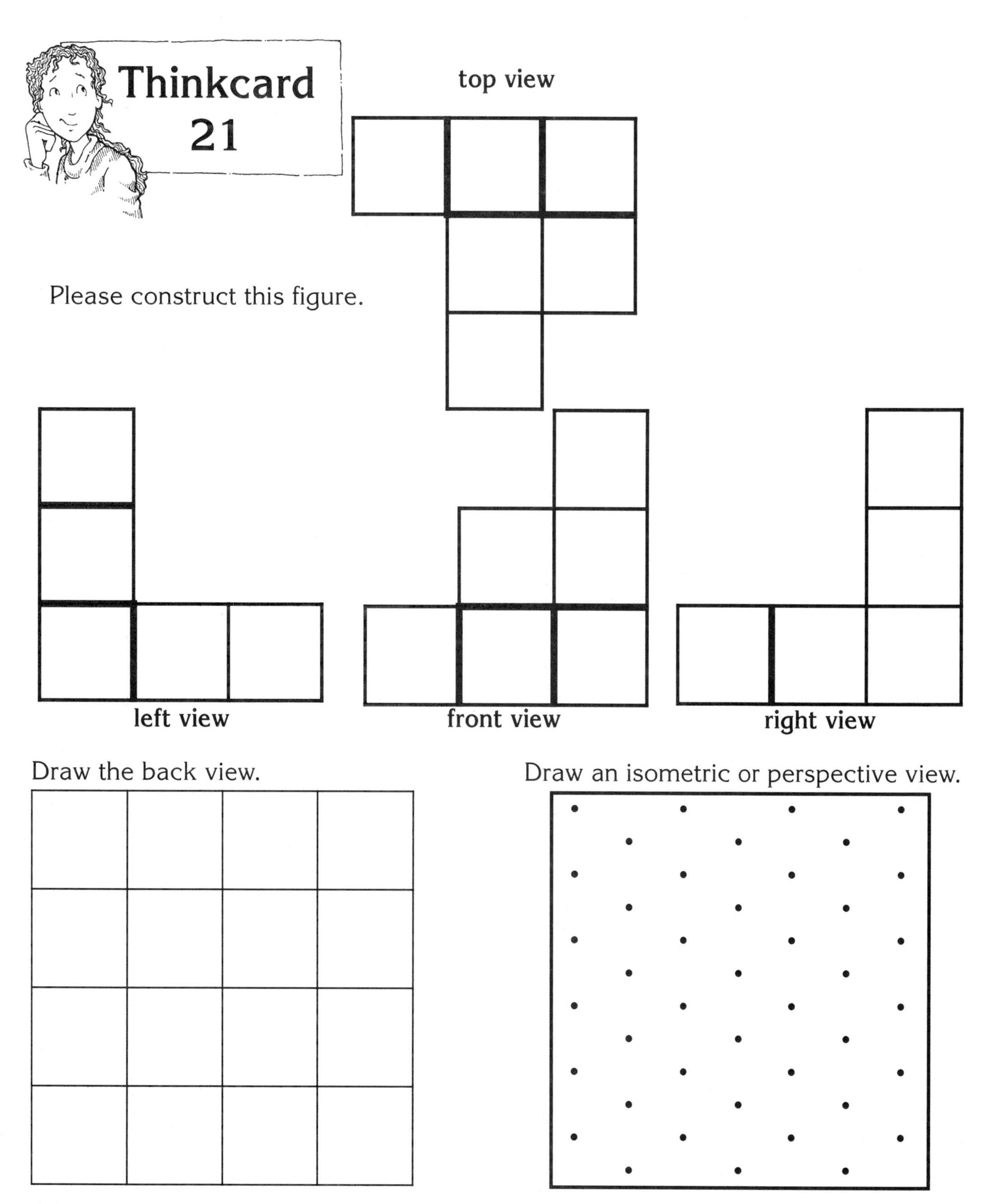

Please construct this figure.

Draw the back view.

Draw an isometric or perspective view.

1. What is the volume?

2. What is the perimeter of the base?

3. What is the total surface area?

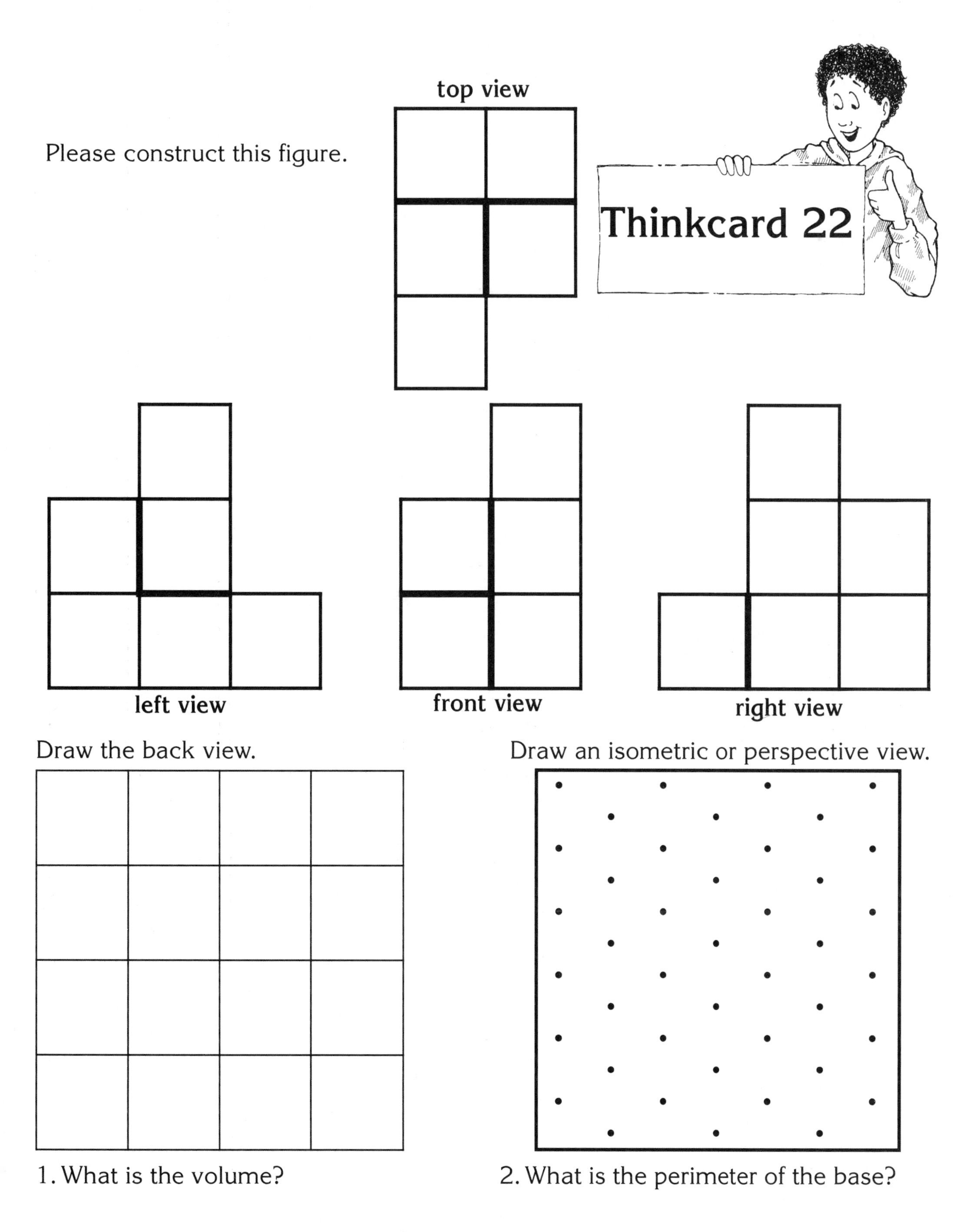

1. What is the volume?

2. What is the perimeter of the base?

3. What is the total surface area?

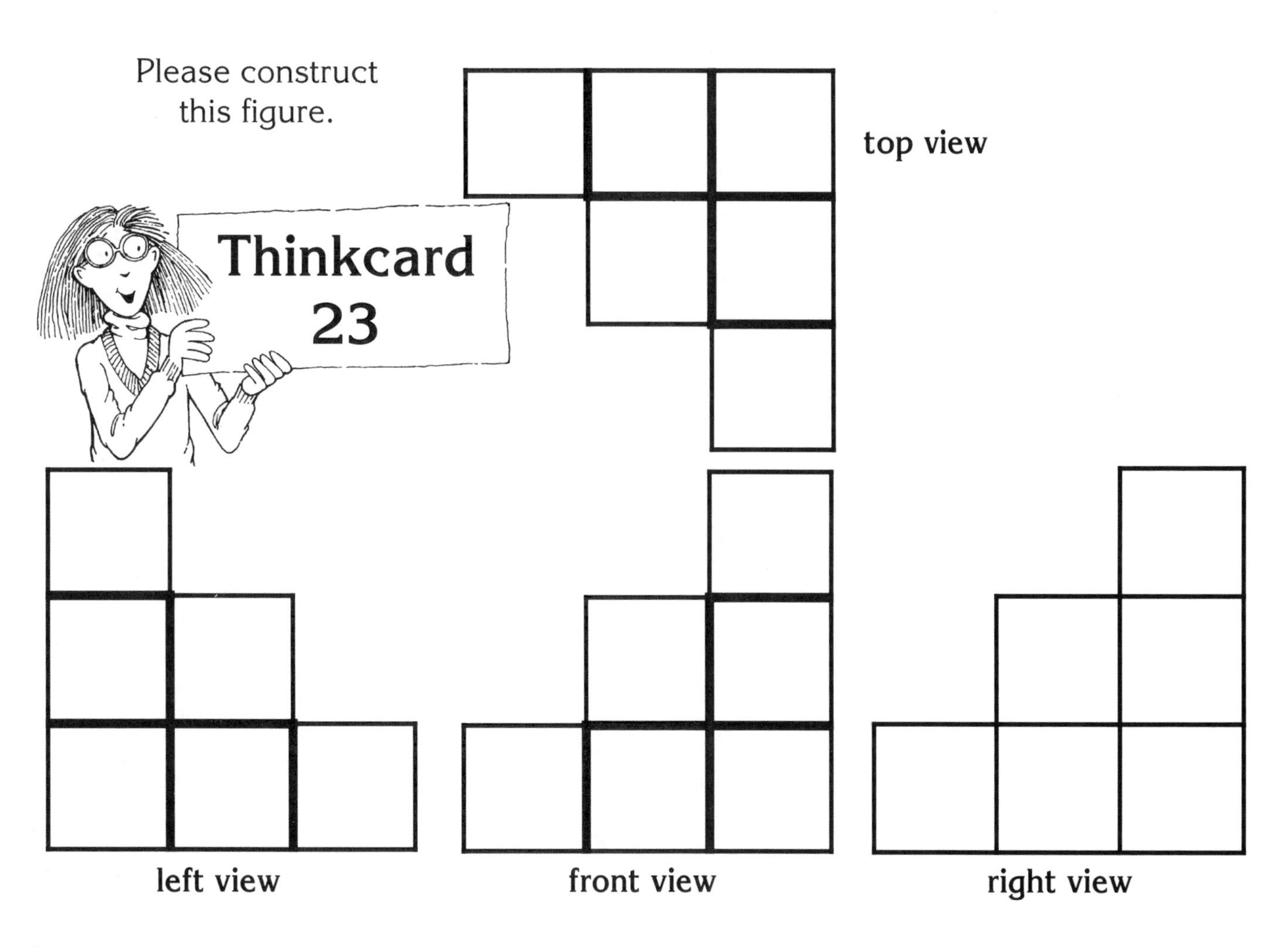

Draw the back view.

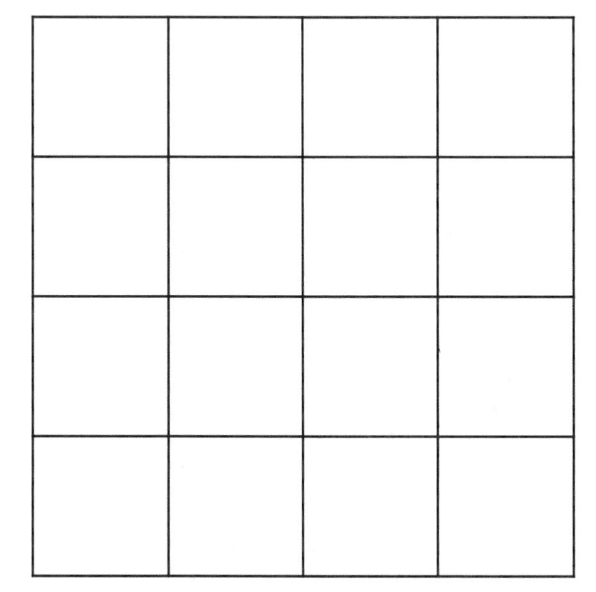

Draw an isometric or perspective view.

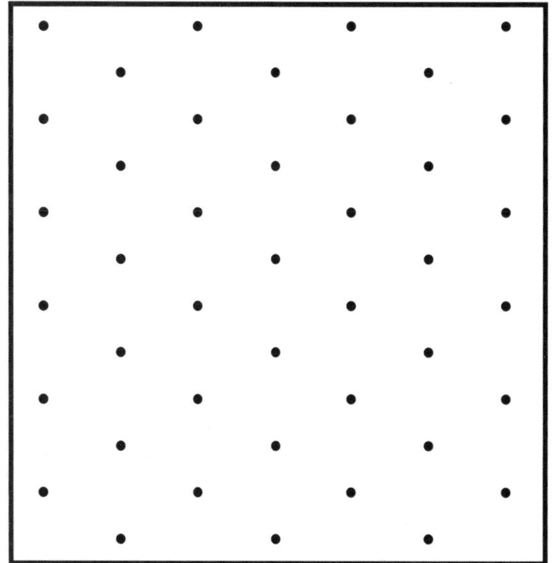

1. What is the volume?

2. What is the perimeter of the base?

3. What is the total surface area?

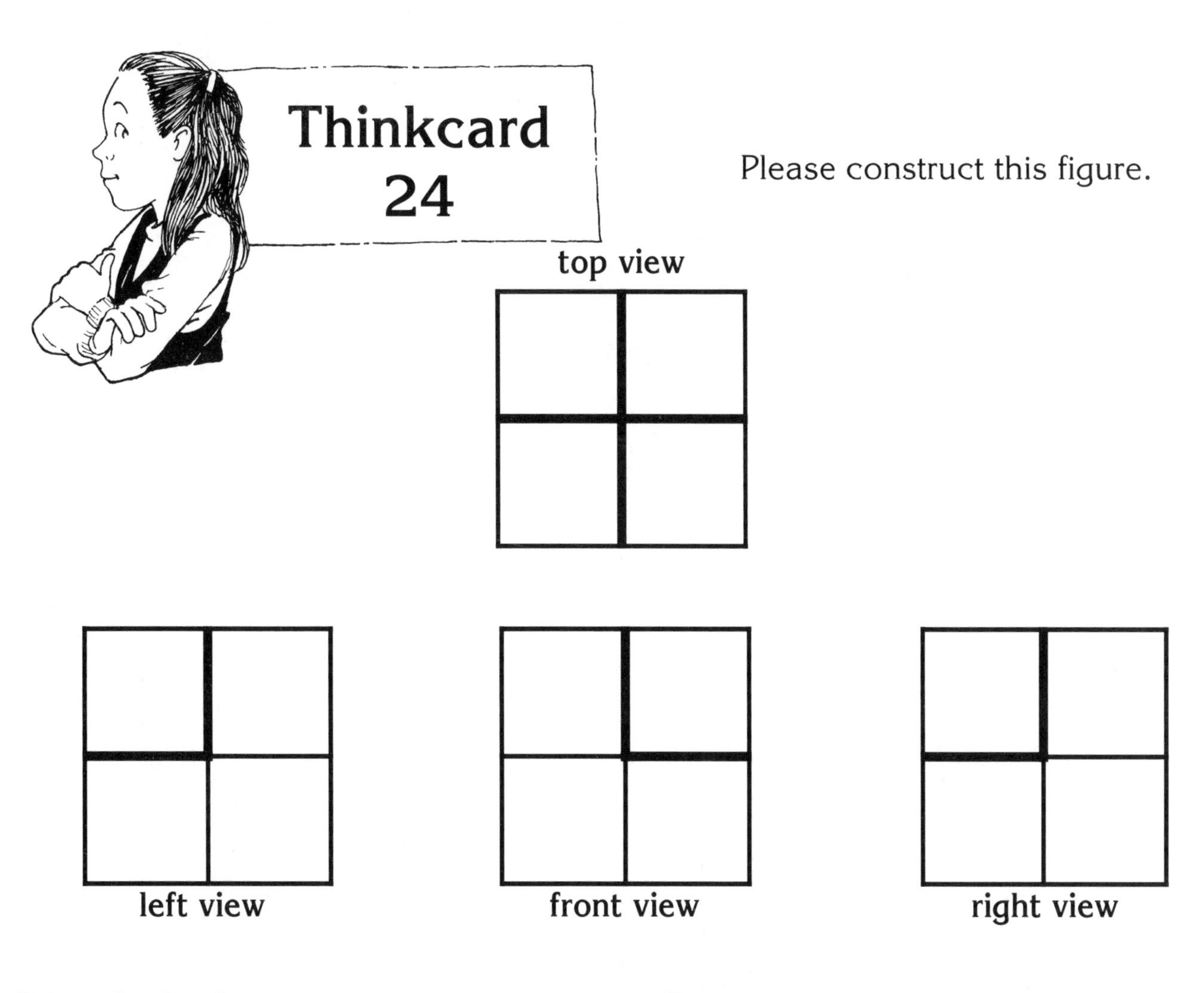

Please construct this figure.

Draw the back view.

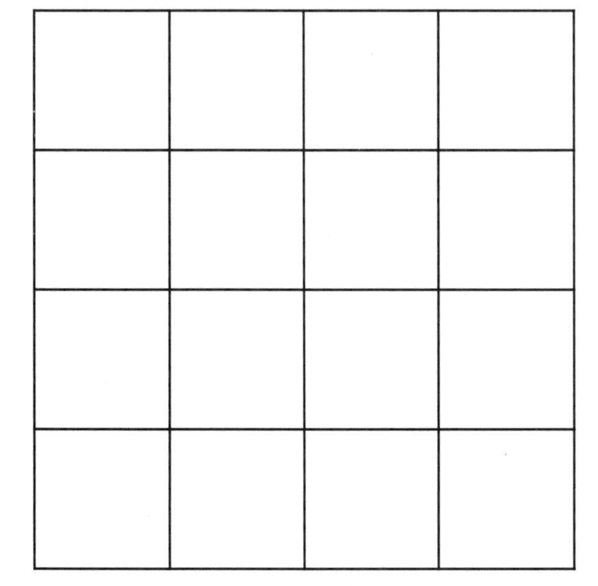

Draw an isometric or perspective view.

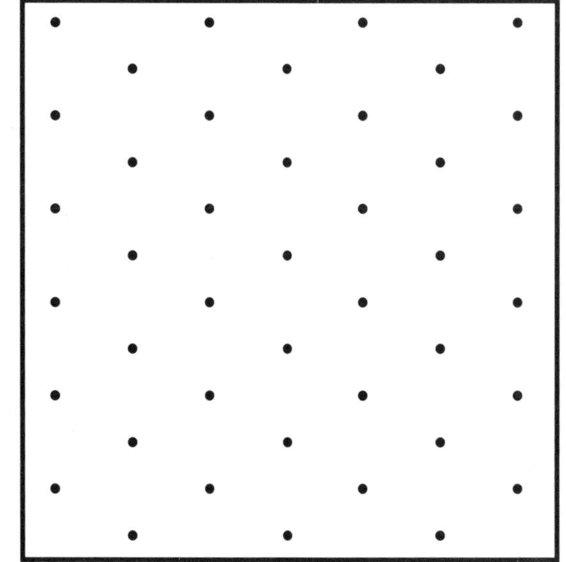

1. What is the volume?

2. What is the perimeter of the base?

3. What is the total surface area?

Spatial Visualization

Part Four: Interpreting Isometric Drawings and Drawing Exploded Views

After students become proficient in drawing isometric views of more complex figures, they are ready to move to the next step of drawing "exploded" views in which an isometric representation of each layer is drawn and corresponding points are connected with broken lines.

Exploded views require students to visualize and make a drawing of each layer separately. They also must determine where key points of each layer meet when joined so they can connect those points with broken lines.

Models in *Part One* and *Part Two* constitute good starting points. Again, it is advisable to tutor the full class and then help individual students as needed. It is also helpful for students to work in small groups so they can discuss steps and help each other.

The solutions for *Thinkcards 25-26* are given below.

Thinkcard 25

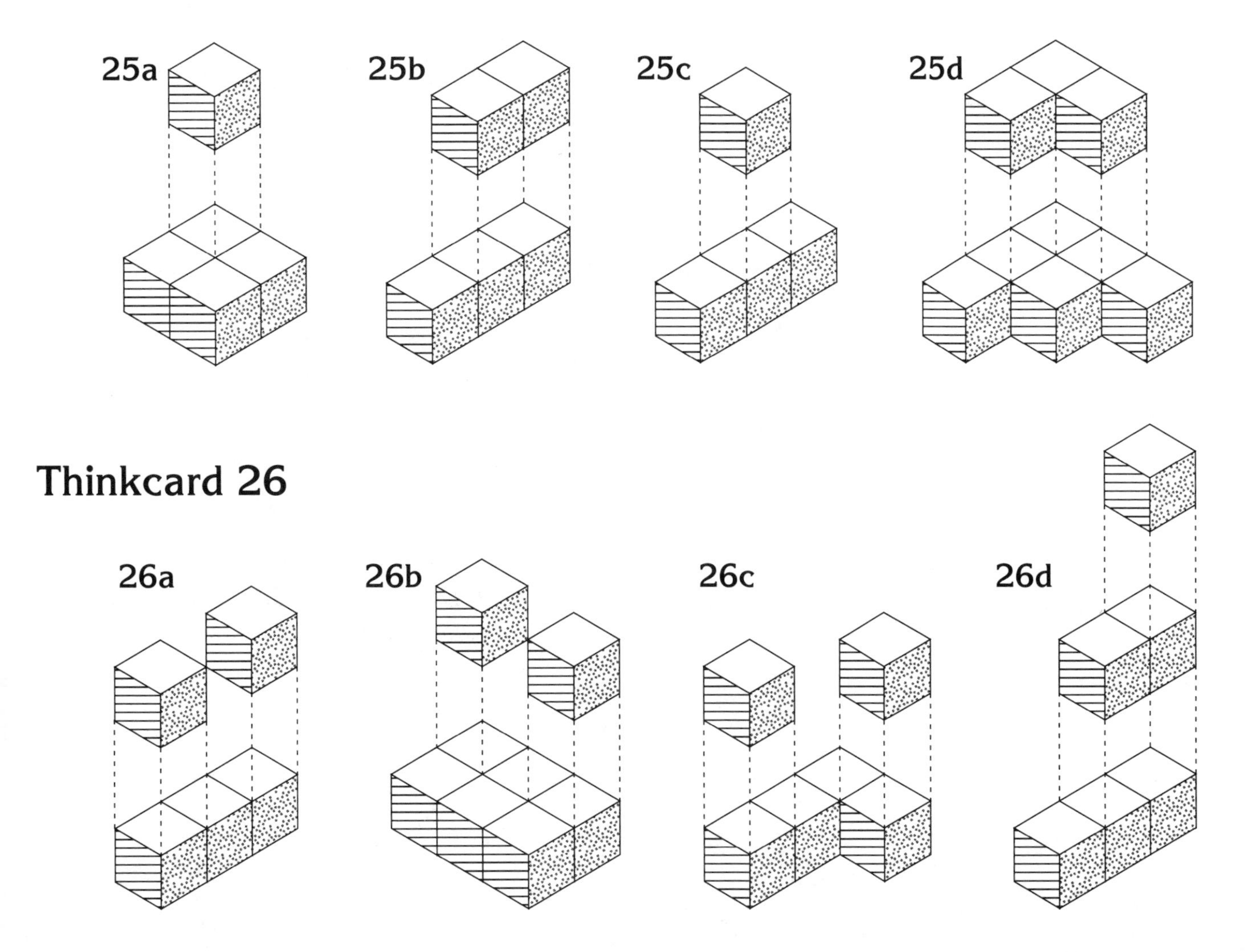

Thinkcards 27-36

The activities on *Thinkcards 27-36* are more comprehensive. Students begin by constructing the model from four-view plans, making the translation from two to three dimensions. The isometric drawings are made by studying the model. This is a translation from three to two dimensions. The exploded isometric view can be drawn by referring to the model, to the isometric drawing, or both. The volume, total surface area, and perimeter of the base are can be determined by making reference to any of the above.

The isometric drawings should use the left-front or front-right orientation, whichever the student chooses. Space is provided on student pages to record the orientation in the lower right- and left-hand corners of the drawing. The arrangement is such that drawings need not be centered over the intersection of left-front or front-right. In each instance, the orientation should be indicated as shown in the examples below.

The isometric and exploded isometric drawings are shown below along with the answers.

Thinkcard 27

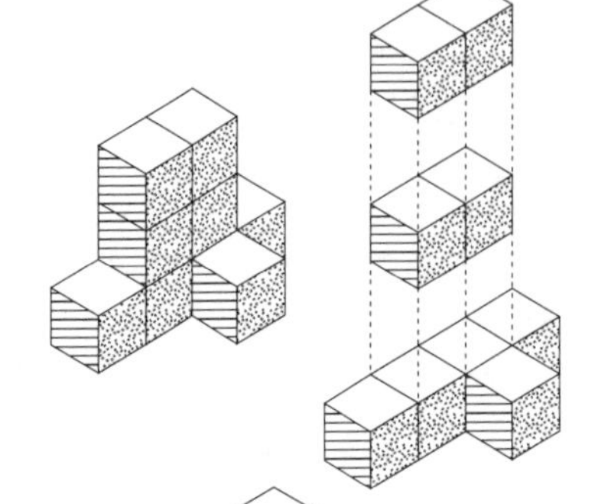

1. Volume: 9 cubic units
2. Surface area: 34 square units
3. Base perimeter: 12 units

Thinkcard 28

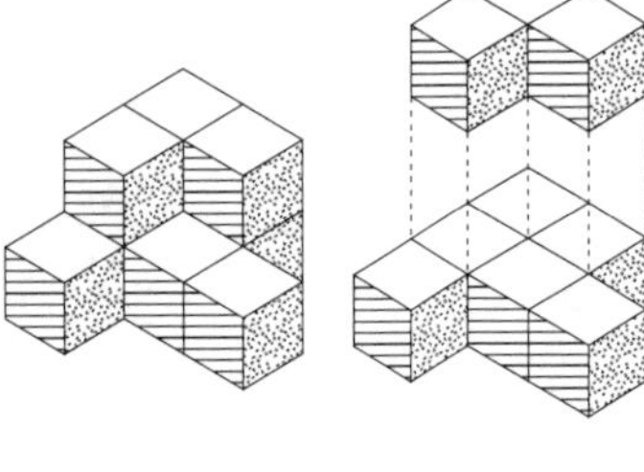

1. Volume: 9 cubic units
2. Surface area: 32 square units
3. Base perimeter: 12 units

Thinkcard 29

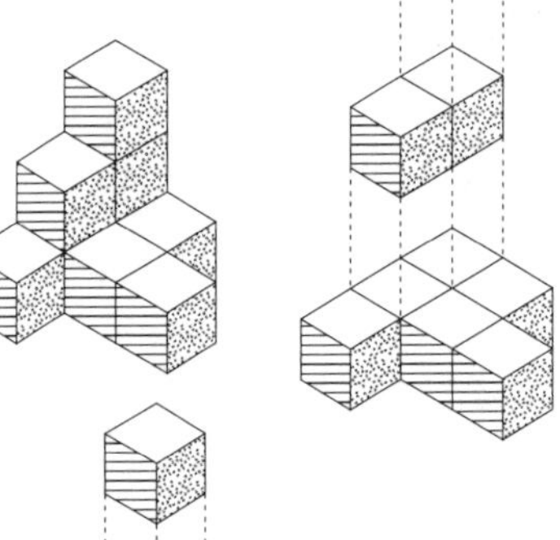

1. Volume: 9 cubic units
2. Surface area: 34 square units
3. Base perimeter: 12 units

Thinkcard 30

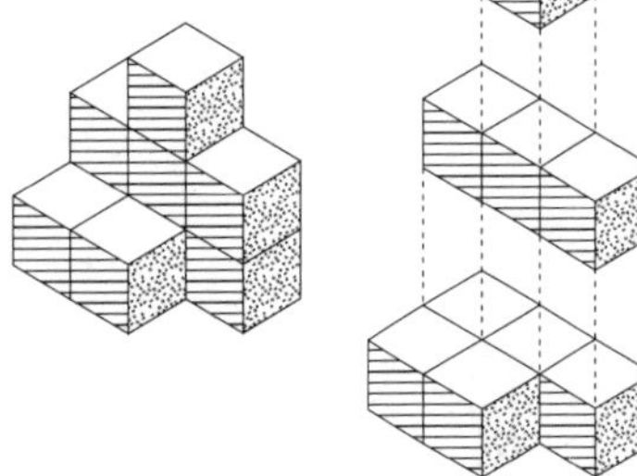

1. Volume: 9 cubic units
2. Surface area: 32 square units
3. Base perimeter: 10 units

Thinkcard 31

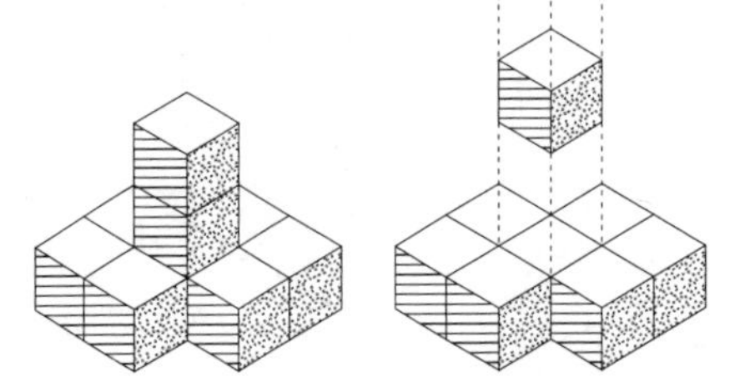

1. Volume: 9 cubic units
2. Surface area: 34 square units
3. Base perimeter: 12 units

Thinkcard 32

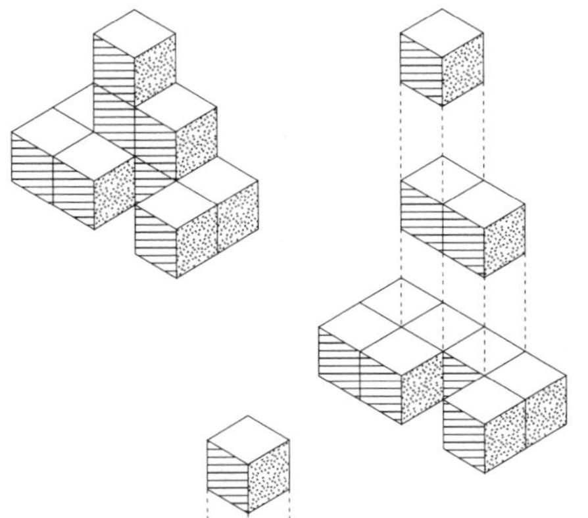

1. Volume: 10 cubic units
2. Surface area: 38 square units
3. Base perimeter: 14 units

Thinkcard 33

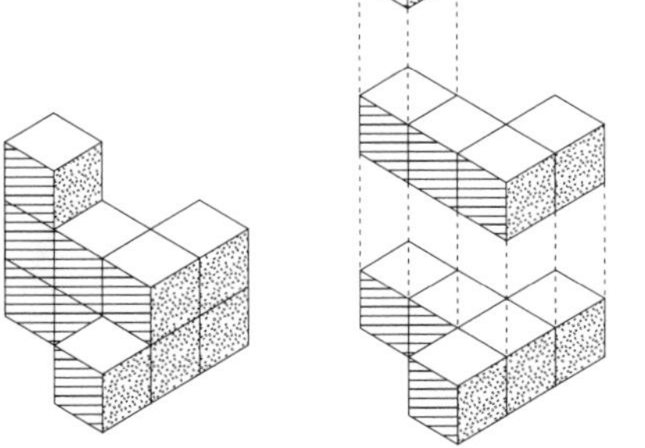

1. Volume: 10 cubic units
2. Surface area: 36 square units
3. Base perimeter: 12 units

Thinkcard 34

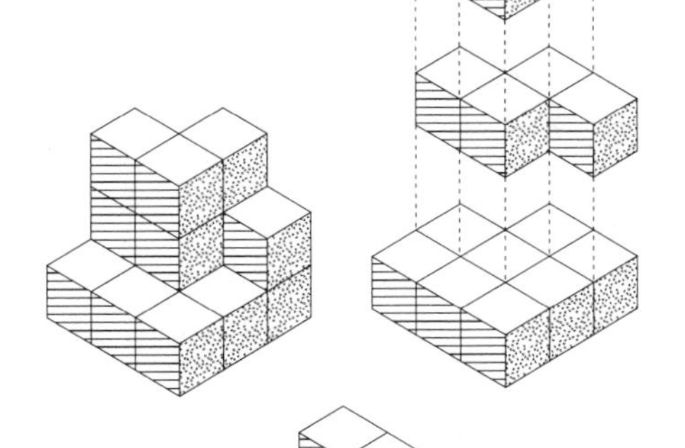

1. Volume: 15 cubic units
2. Surface area: 46 square units
3. Base perimeter: 12 units

Thinkcard 35

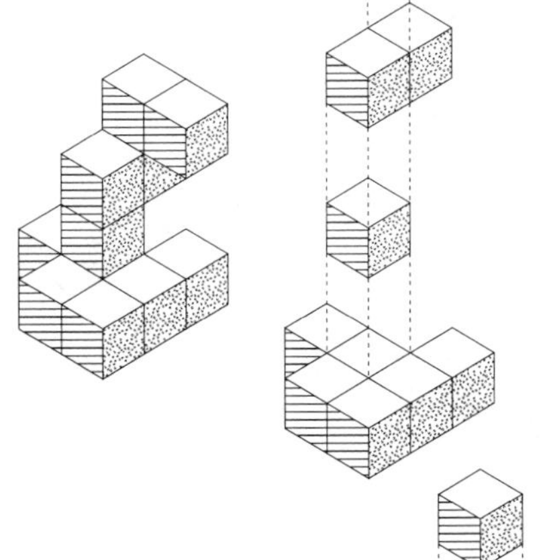

1. Volume: 11 cubic units
2. Surface area: 44 square units
3. Base perimeter: 12 units

Thinkcard 36

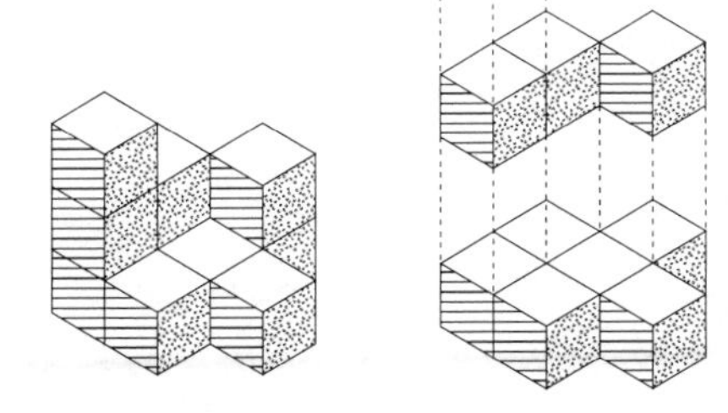

1. Volume: 10 cubic units
2. Surface area: 38 square units
3. Base perimeter: 12 units

Thinkcard 25

Please build each of these models and then draw an exploded isometric view showing the layers of each.

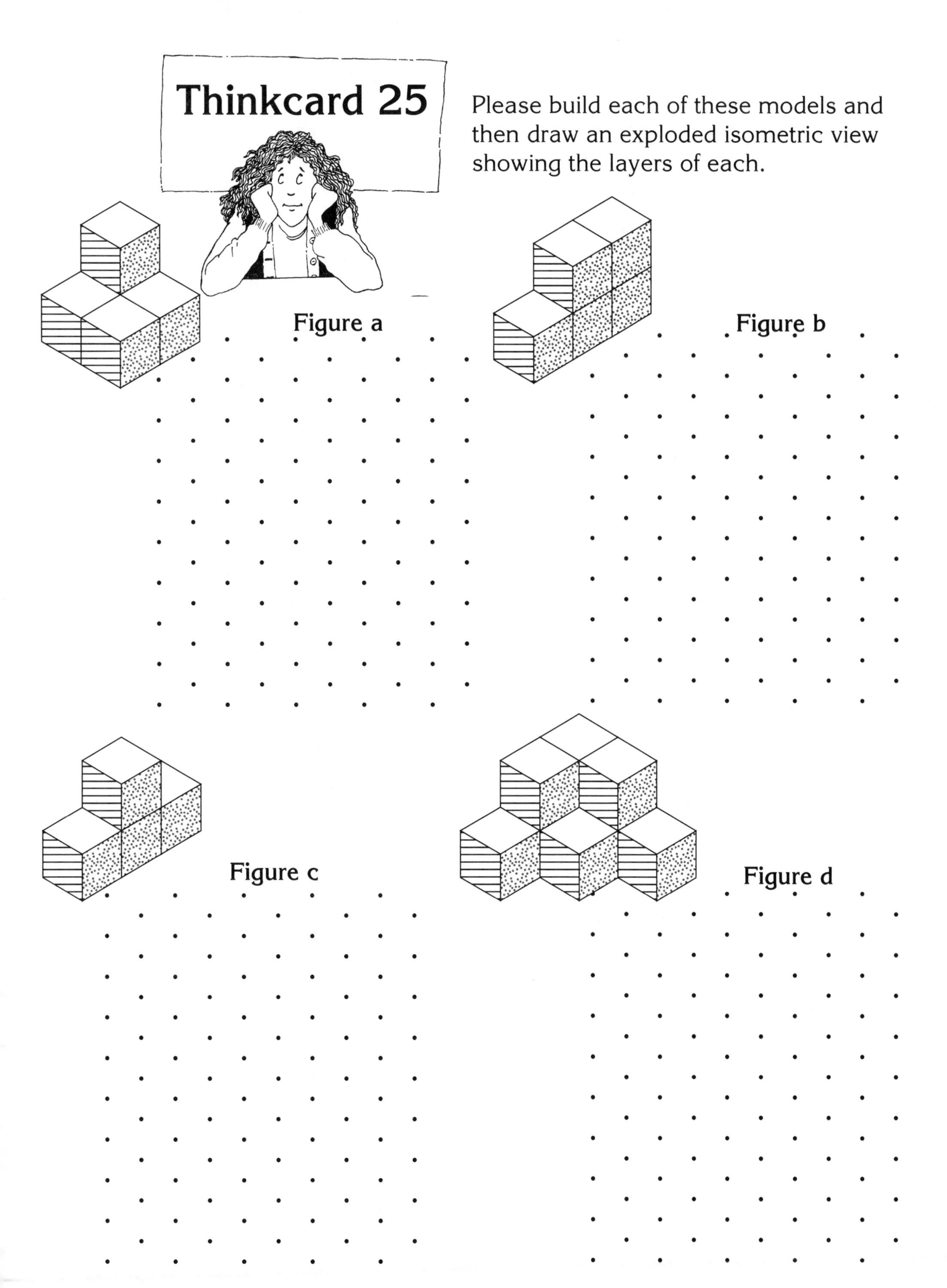

Thinkcard 26

Please build each of these models and then draw an exploded isometric view showing the layers of each.

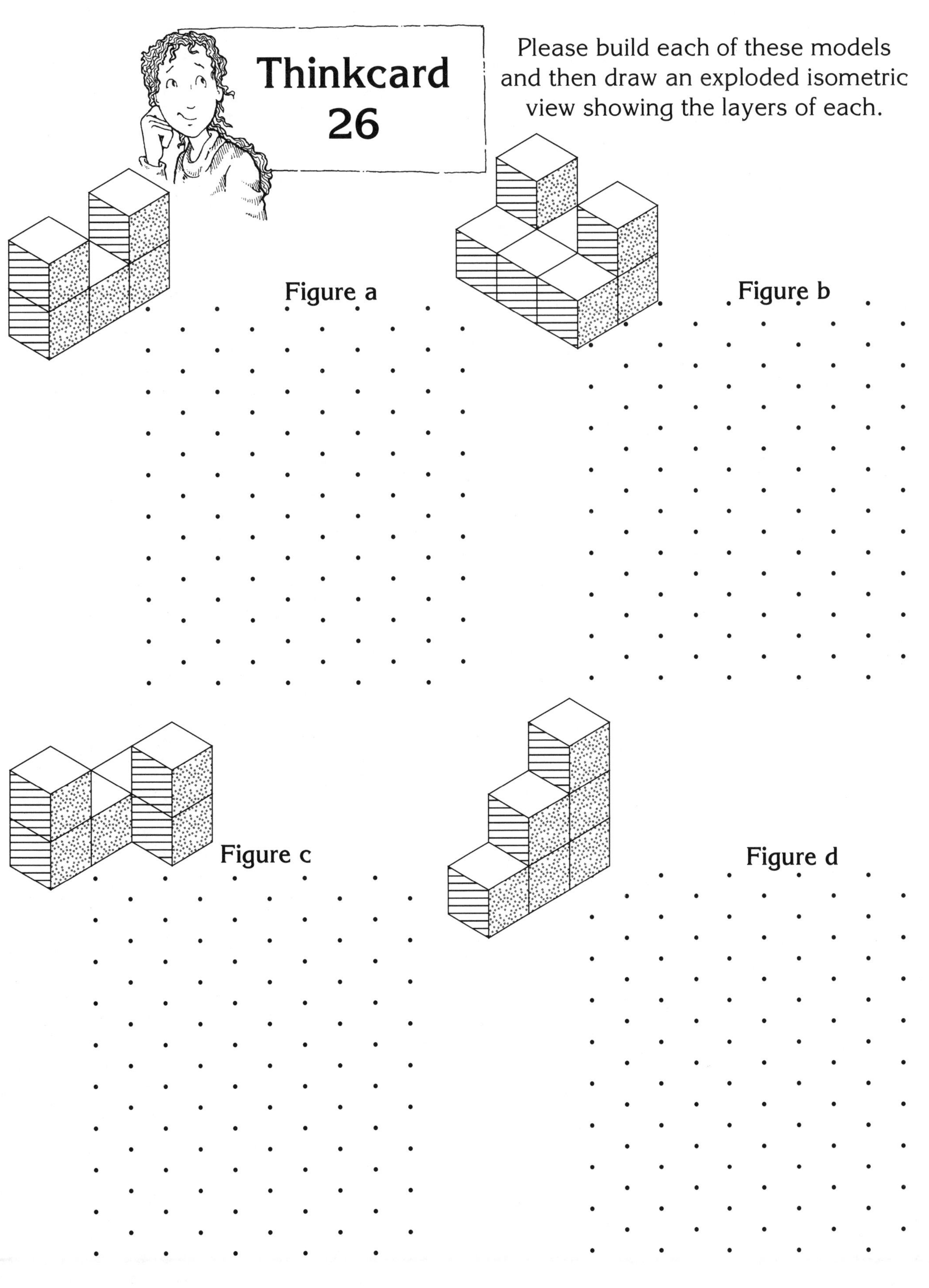

Thinkcard 27

Please construct this figure. The edge of one block equals one unit of length. Use a left-front or front-right orientation in making the isometric drawings. Label the orientation.

1. What is the volume?

2. What is the total surface area?

3. What is the perimeter of the base?

Please construct this figure. The edge of one block equals one unit of length. Use a left-front or front-right orientation in making the isometric drawings. Label the orientation.

1. What is the volume?

2. What is the total surface area?

3. What is the perimeter of the base?

1. What is the volume?

2. What is the total surface area?

3. What is the perimeter of the base?

Please construct this figure. The edge of one block equals one unit of length. Use a left-front or front-right orientation in making the isometric drawings. Label the orientation.

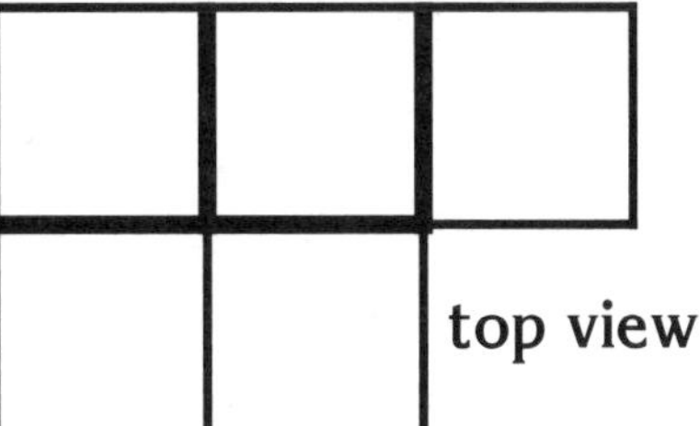

top view

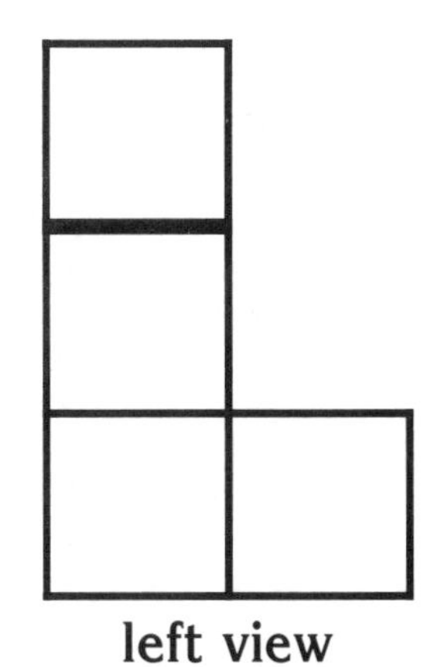

left view

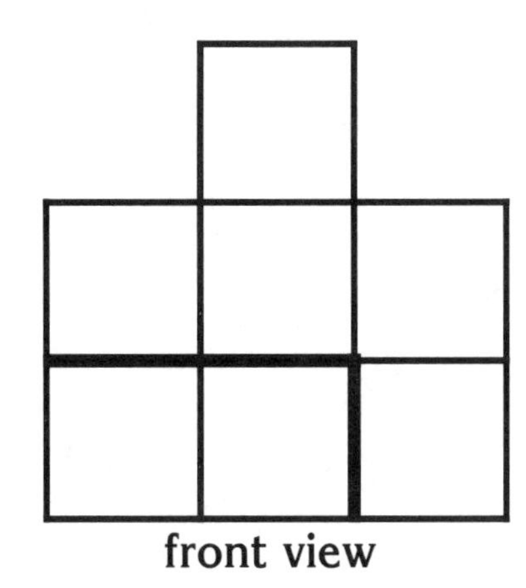

front view

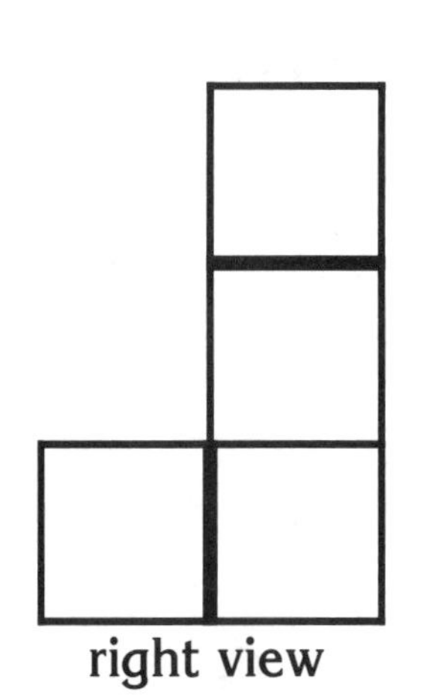

right view

Draw an isometric view.

Draw an exploded isometric view.

1. What is the volume?

2. What is the total surface area?

3. What is the perimeter of the base?

Please construct this figure. The edge of one block equals one unit of length. Use a left-front or front-right orientation in making the isometric drawings. Label the orientation.

Thinkcard 31

top view

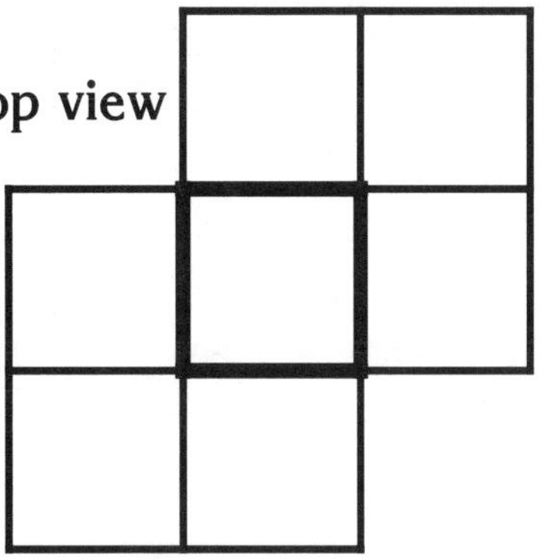

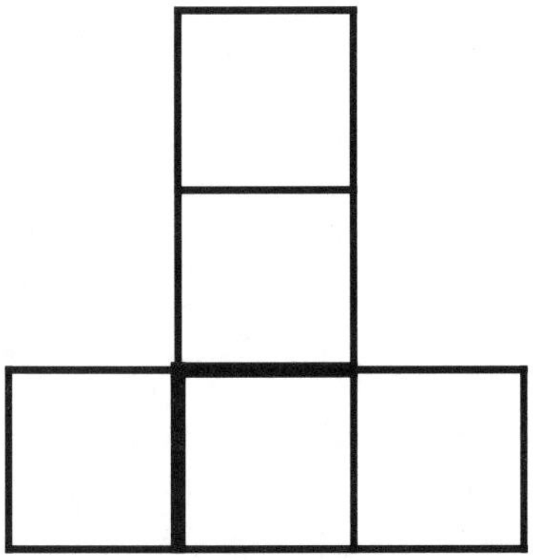

left view

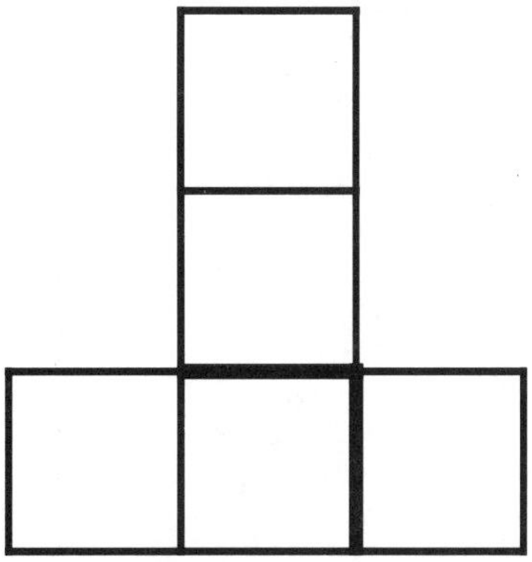

front view

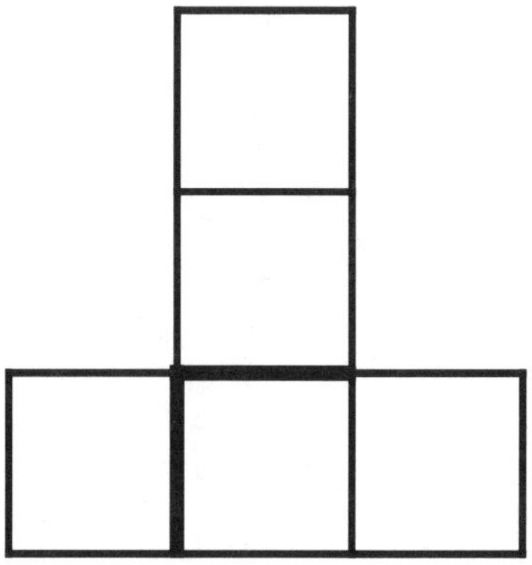

right view

Draw an isometric view.

Draw an exploded isometric view.

1. What is the volume?

2. What is the total surface area?

3. What is the perimeter of the base?

Please construct this figure. The edge of one block equals one unit of length. Use a left-front or front-right orientation in making the isometric drawings. Label the orientation.

Thinkcard 32

1. What is the volume?

2. What is the total surface area?

3. What is the perimeter of the base?

1. What is the volume?

2. What is the total surface area?

3. What is the perimeter of the base?

Thinkcard 34

Please construct this figure. The edge of one block equals one unit of length. Use a left-front or front-right orientation in making the isometric drawings. Label the orientation.

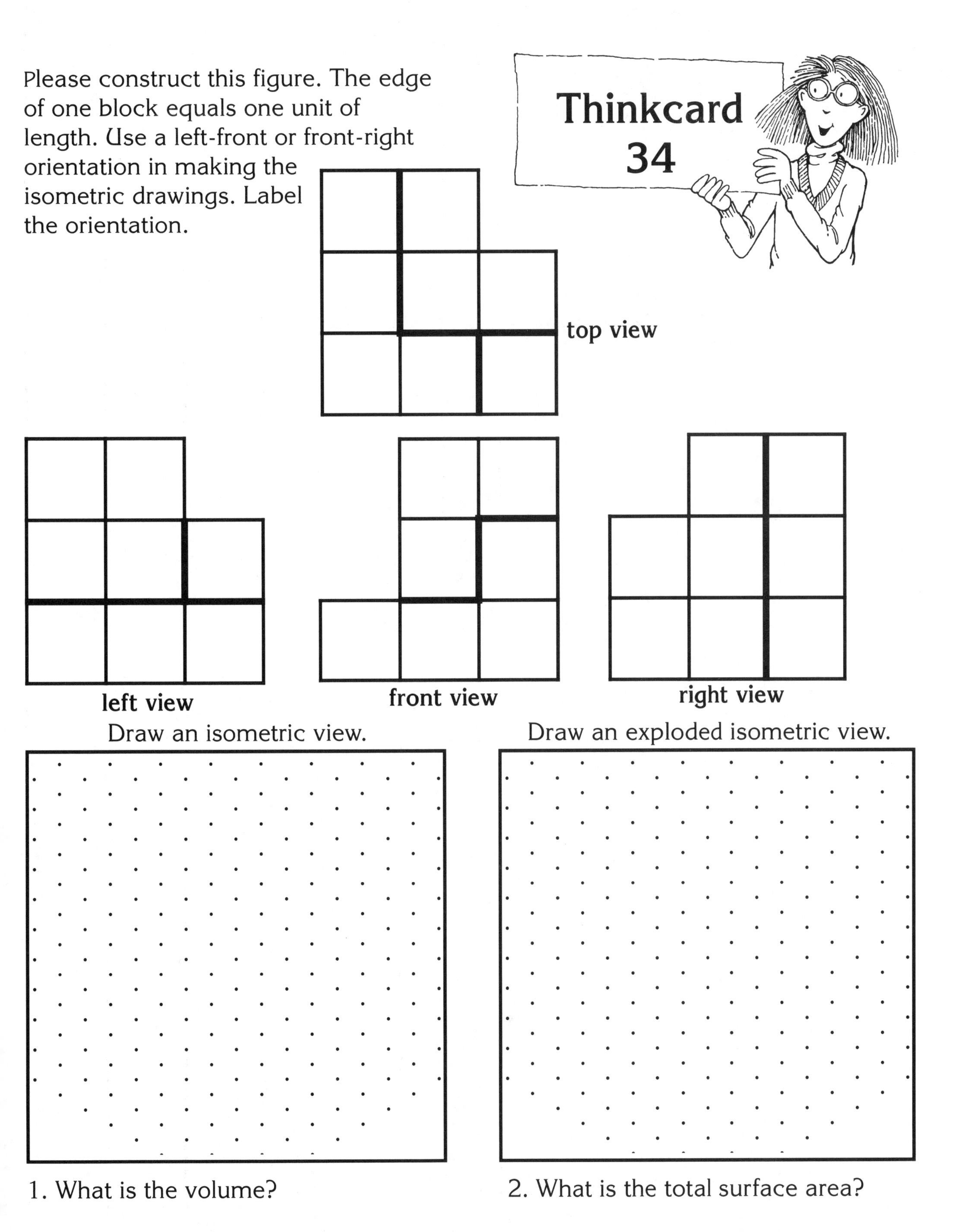

1. What is the volume?

2. What is the total surface area?

3. What is the perimeter of the base?

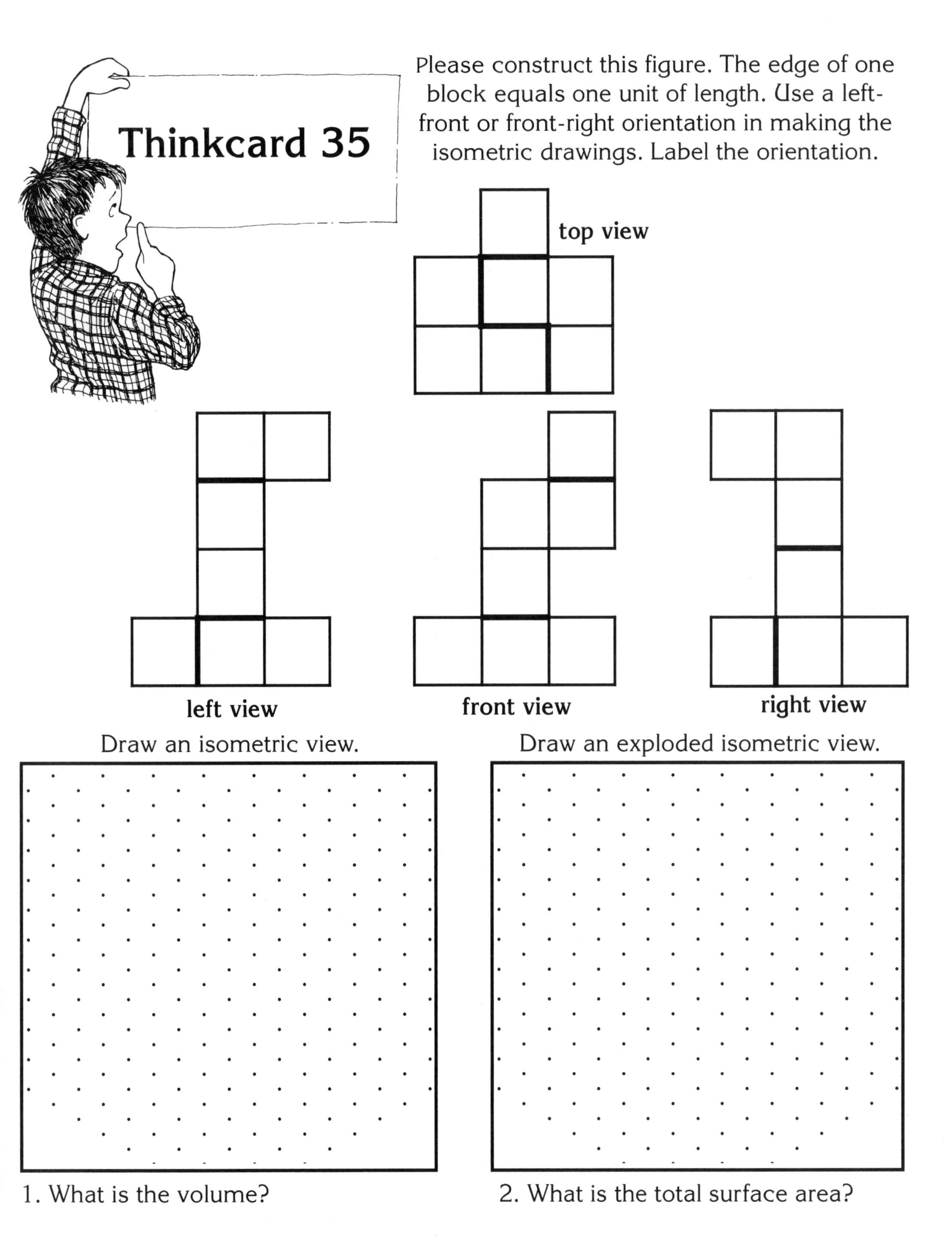

Thinkcard 35

Please construct this figure. The edge of one block equals one unit of length. Use a left-front or front-right orientation in making the isometric drawings. Label the orientation.

Draw an isometric view.

Draw an exploded isometric view.

1. What is the volume?

2. What is the total surface area?

3. What is the perimeter of the base?

Please construct this figure. The edge of one block equals one unit of length. Use a left-front or front-right orientation in making the isometric drawings. Label the orientation.

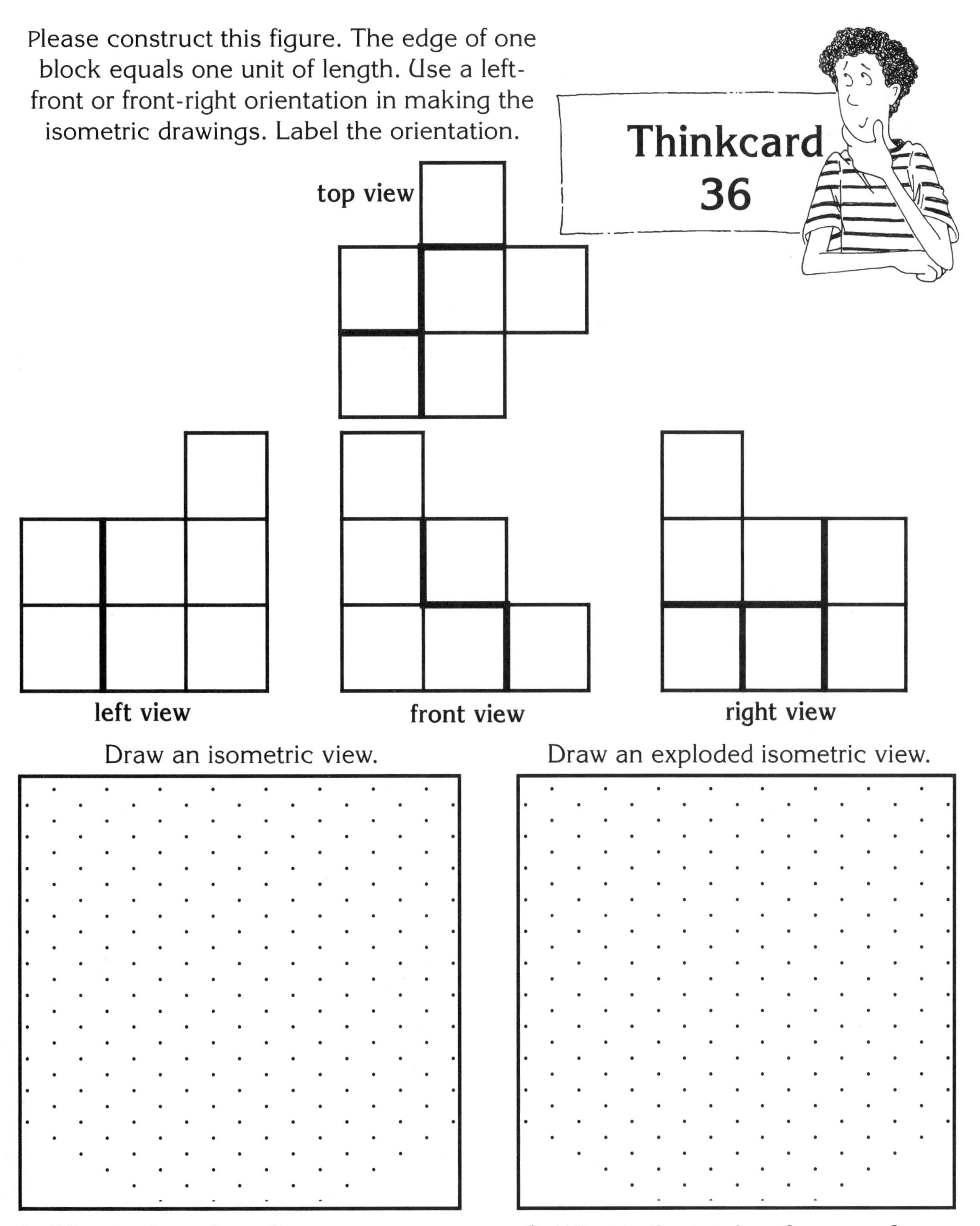

1. What is the volume?

2. What is the total surface area?

3. What is the perimeter of the base?

Spatial Visualization

Part Five: Interpreting Stacks of Blocks

In *Part Five* students are presented with the perspective drawings of stacks of blocks. The three hidden faces are completely smooth as though the stack has been packed into the corner of a box. Since no blocks are missing in the invisible portion, it is possible to determine the volume.

Students first determine the volume, surface area, and base perimeter from the drawing. They will develop their individual techniques for this task. Encourage individual students to first work on their own in the development of these techniques and then to share them, thereby enriching their spatial understanding. In some instances, students may benefit by first constructing a model of the stack as a way to check their computations.

Students complete the activity by making exploded-view isometric drawings of the stack.

Answers

Thinkcard 37

37 A

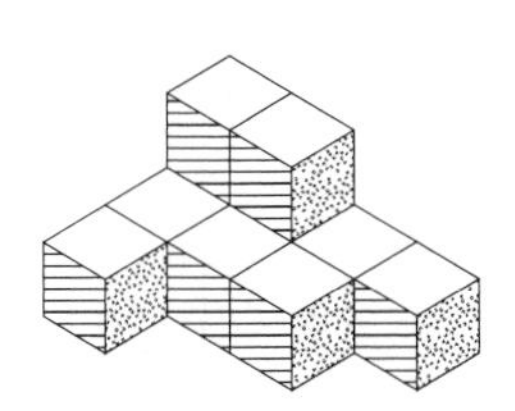

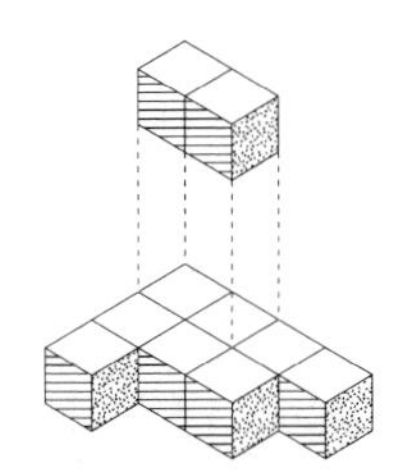

Volume: 10 cubic units
Surface area: 36 square units
Base perimeter: 14 units

37 B

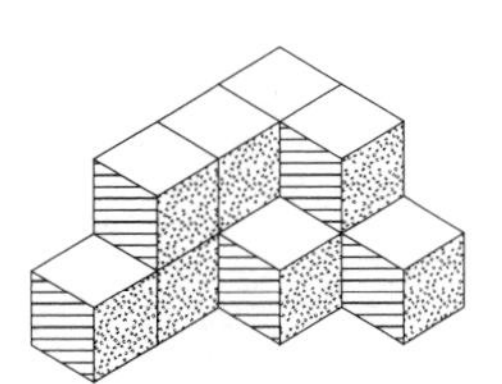

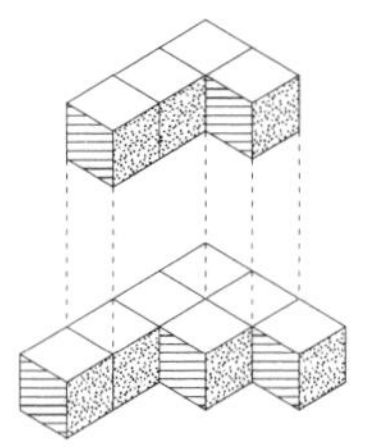

Volume: 11 cubic units
Surface area: 38 square units
Base perimeter: 14 units

Thinkcard 38

38 A

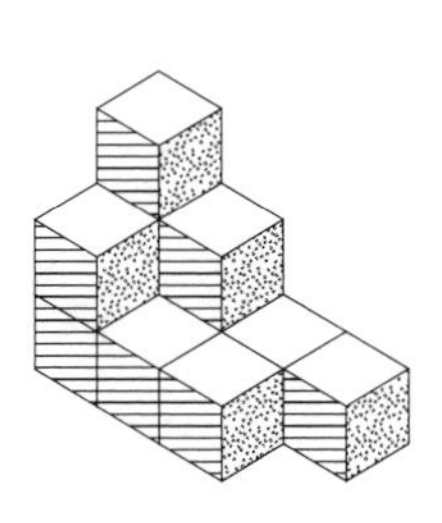

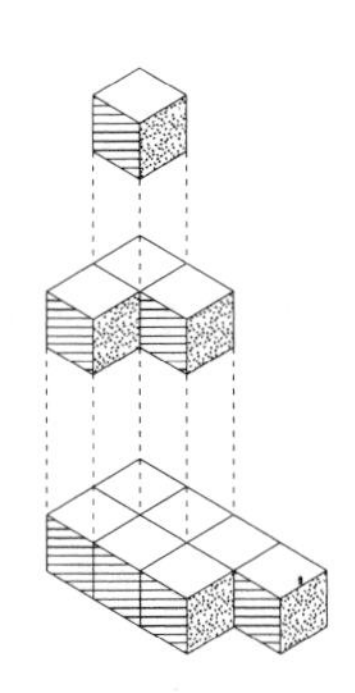

Volume: 11 cubic units
Surface area: 38 square units
Base perimeter: 12 units

38 B

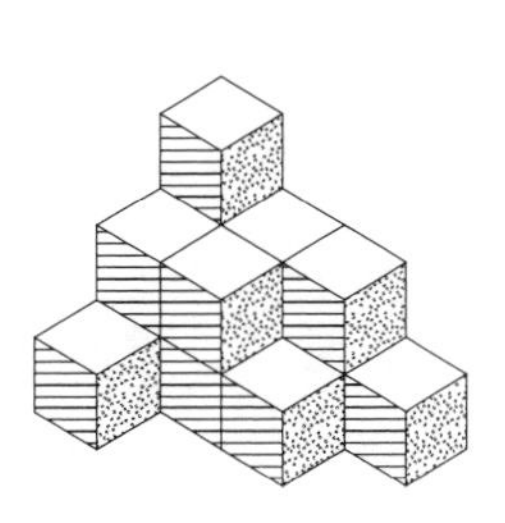

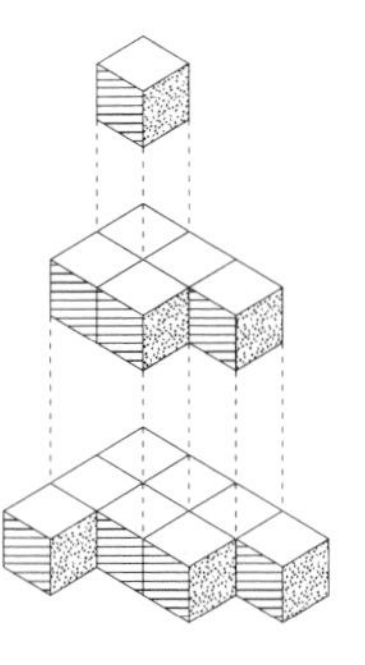

Volume: 14 cubic units
Surface area: 44 square units
Base perimeter: 14 units

Answers

Thinkcard 39

39 A

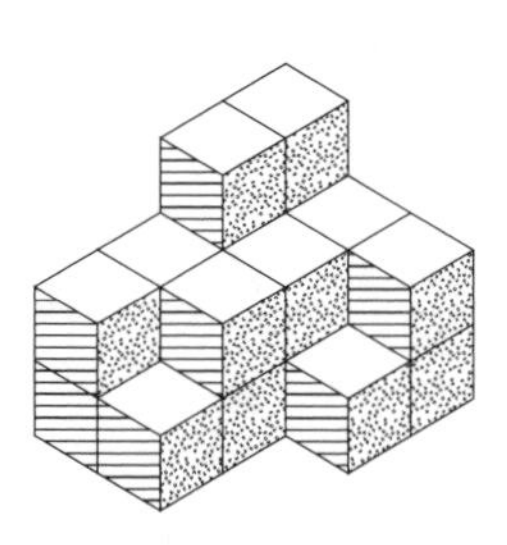

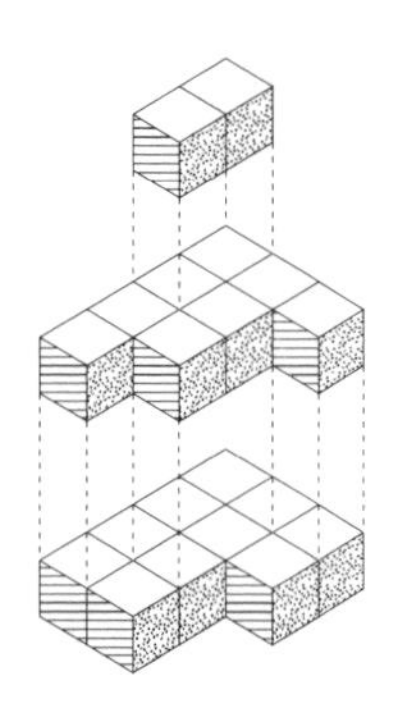

Volume: 20 cubic units
Surface area: 54 square units
Base perimeter: 14 units

39 B

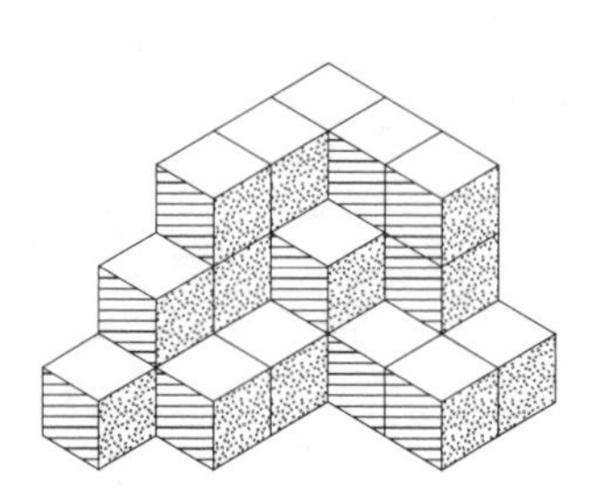

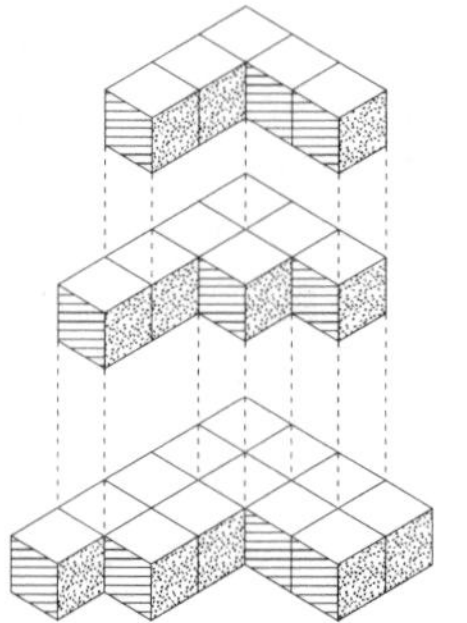

Volume: 25 cubic units
Surface area: 70 square units
Base perimeter: 18 units

Thinkcard 40

40A

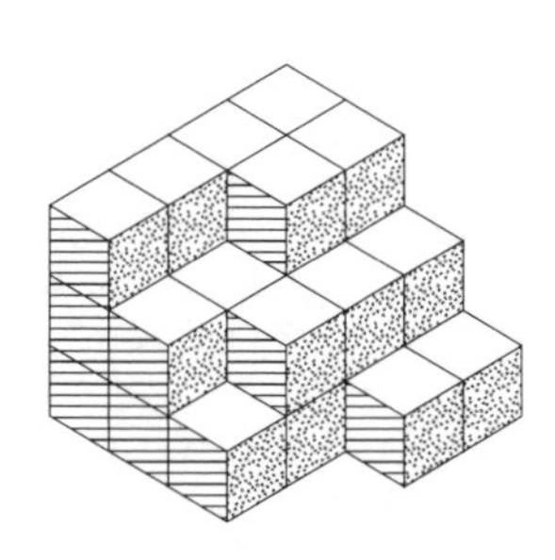

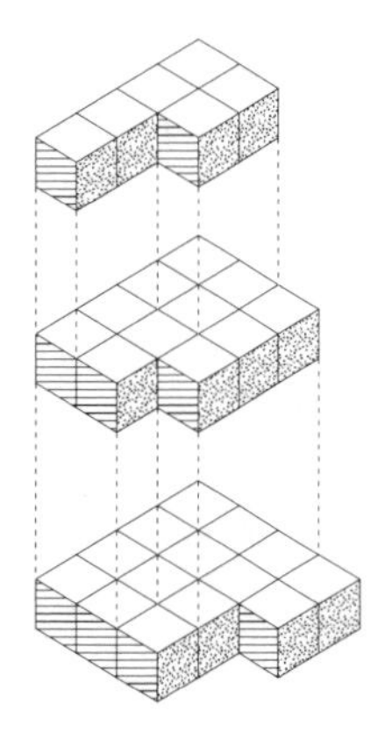

Volume: 31 cubic units
Surface area: 70 square units
Base perimeter: 16 units

40 B

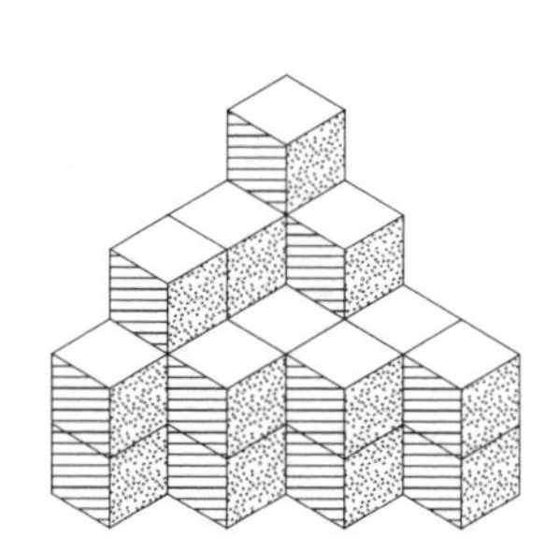

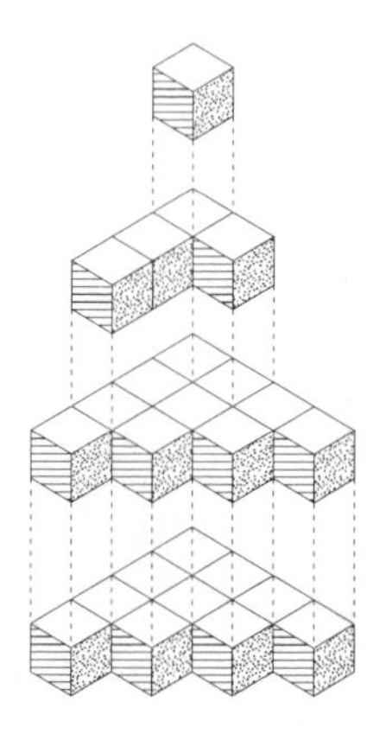

Volume: 25 cubic units
Surface area: 66 square units
Base perimeter: 16 units

Answers

Thinkcard 41

41 A

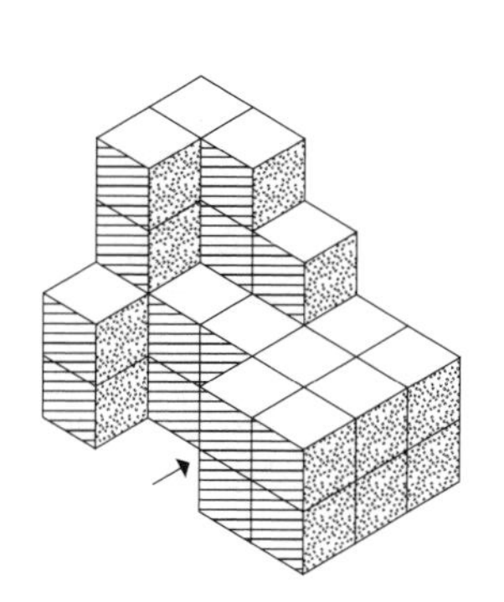

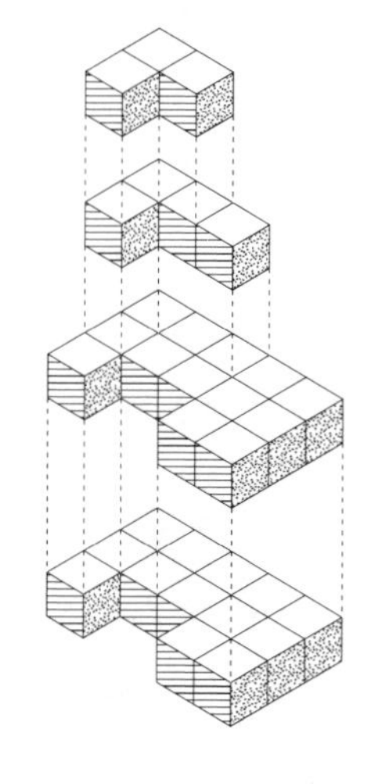

Volume: 33 cubic units
Surface area: 116 square units
Base perimeter: 26 units

41 B

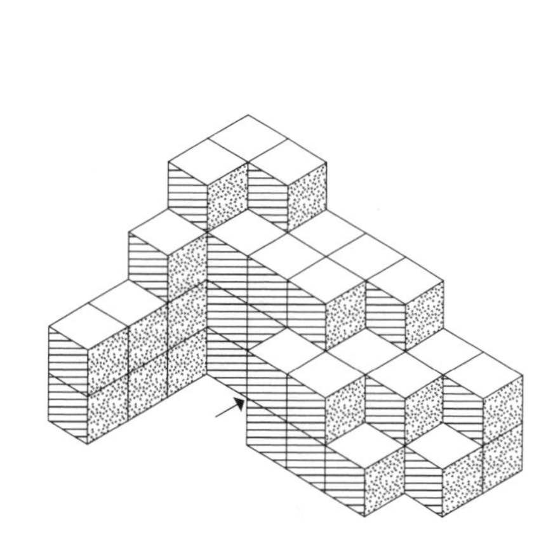

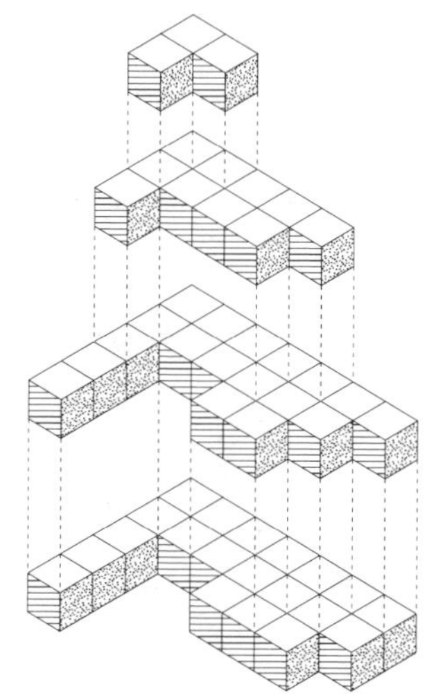

Volume: 51 cubic units
Surface area: 116 square units
Base perimeter: 26 units

Both Figure A and B have hidden faces whose location is indicated by the arrows. These hidden faces impact visualizing the surface area and perimeter. The perimeter is best determined by examining the bottom layer in the exploded drawing. If necessary, the model should be constructed to show where the faces are hidden.

Thinkcard 42

42 A

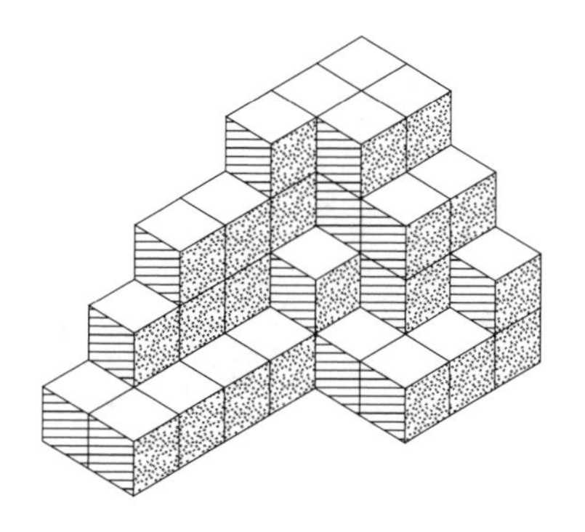

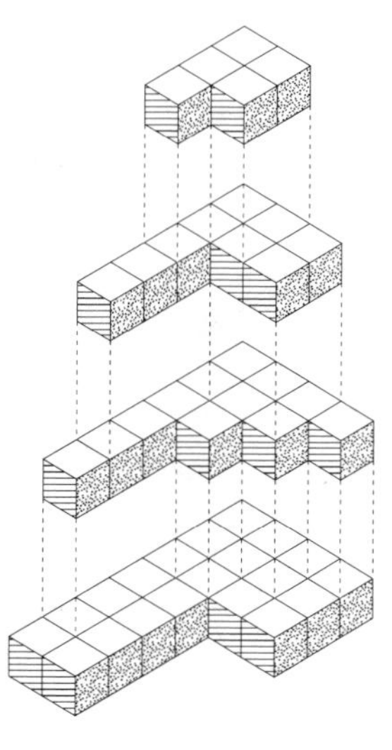

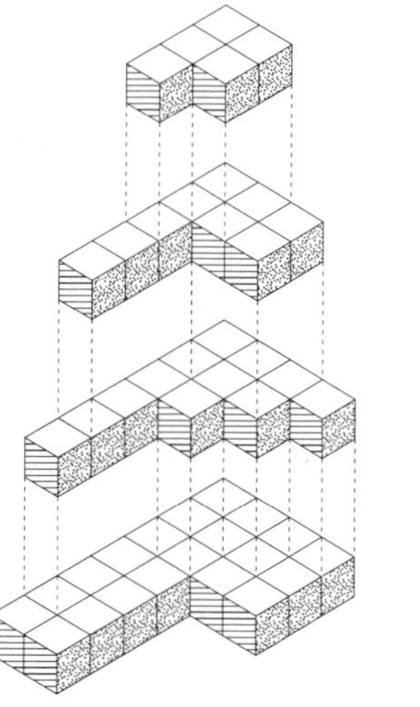

Volume: 46 cubic units
Surface area: 108 square units
Base perimeter: 22 units

42 B

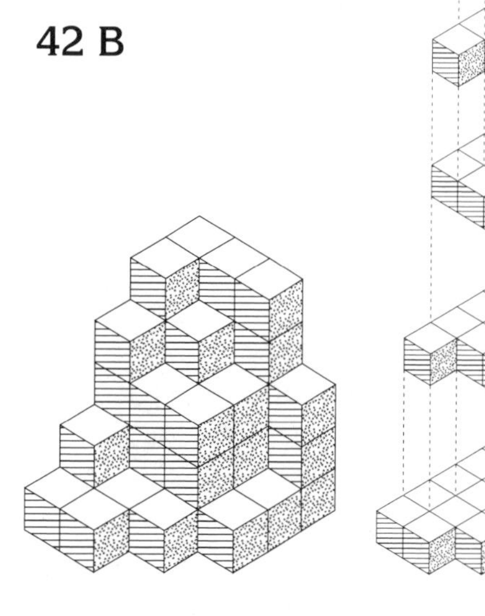

Volume: 48 cubic units
Surface area: 104 square units
Base perimeter: 18 units

Answers

Thinkcard 43

43 A

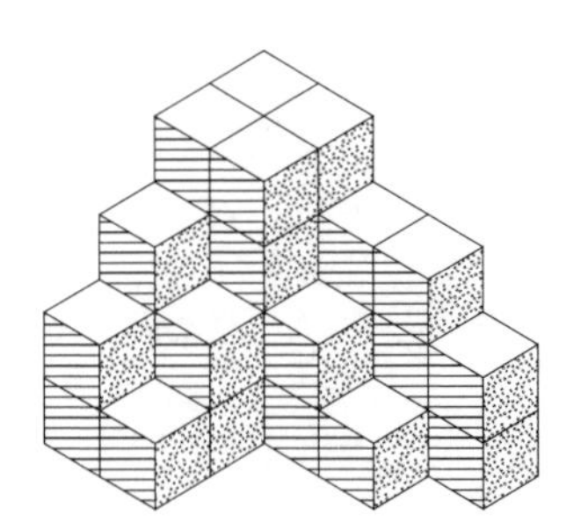

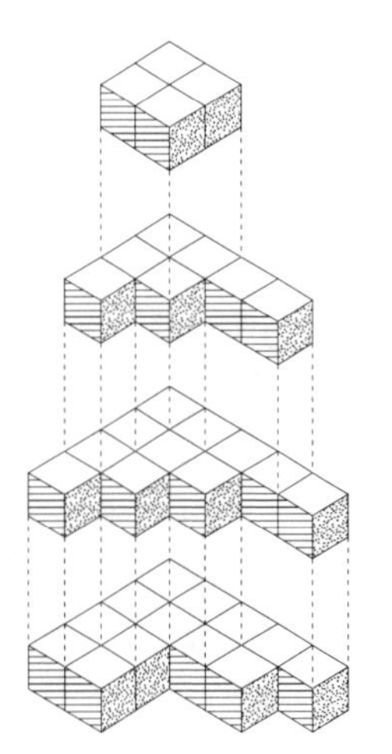

Volume: 35 cubic units
Surface area: 84 square units
Base perimeter: 18 units

43 B

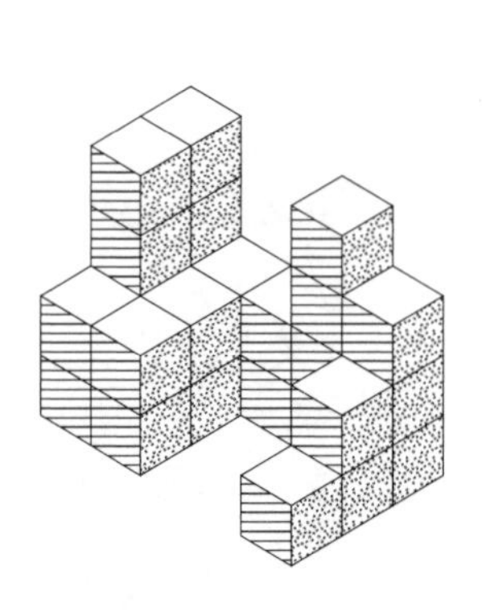

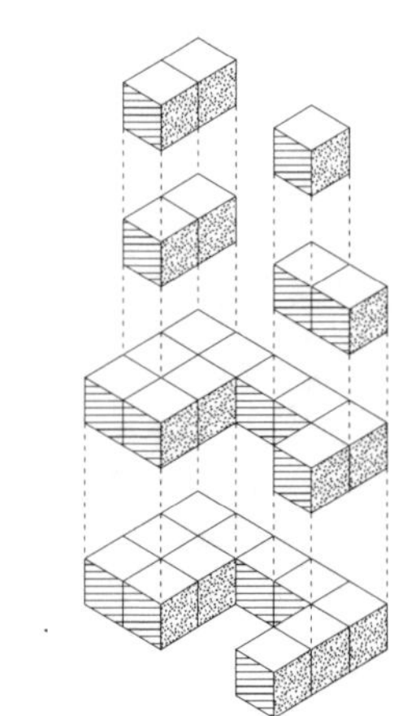

Volume: 28 cubic units
Surface area: 82 square units
Base perimeter: 20 units

Thinkcard 44

44 A

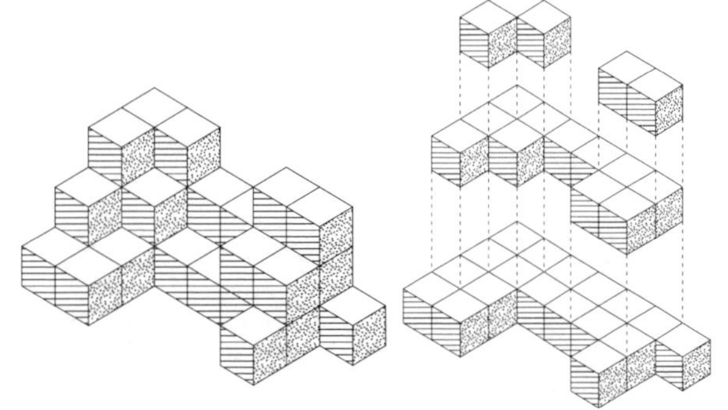

Volume: 10 cubic units
Surface area: 36 square units
Base perimeter: 14 units

44 B

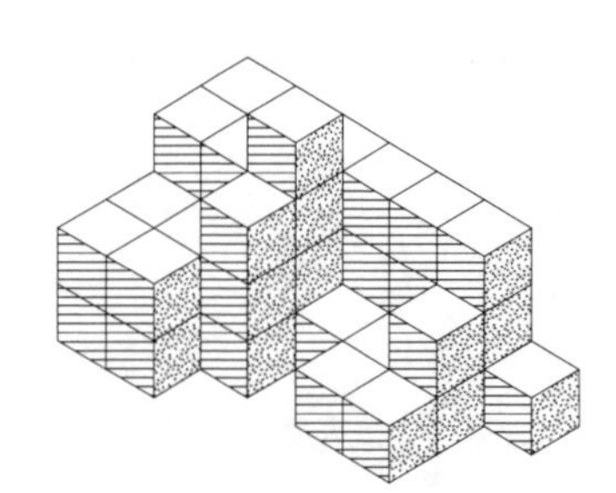

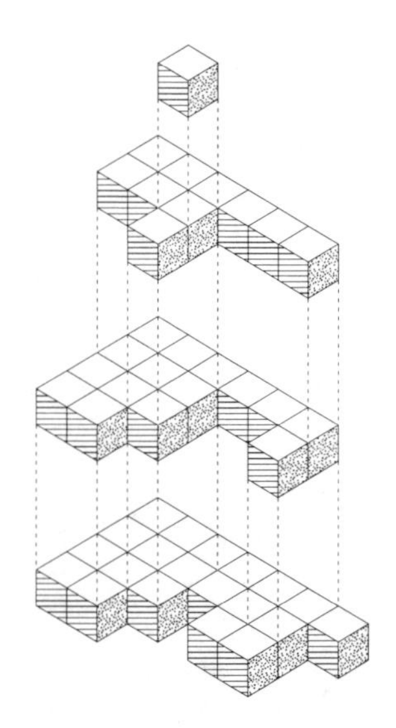

Volume: 46 cubic units
Surface area: 108 square units
Base perimeter: 24 units

Answers

Thinkcard 45

45 A

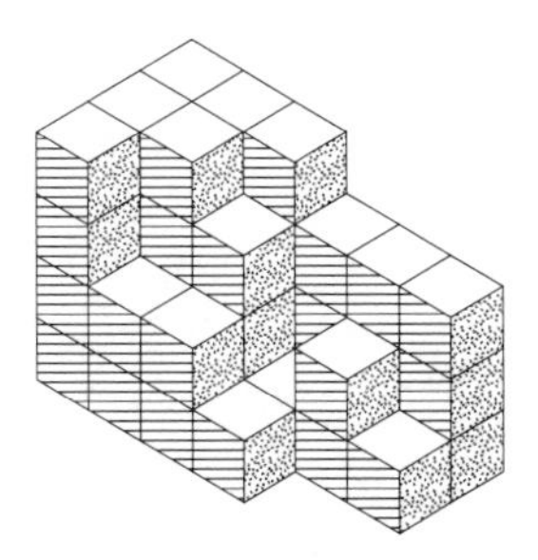

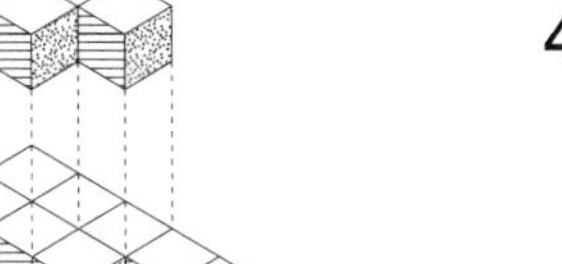

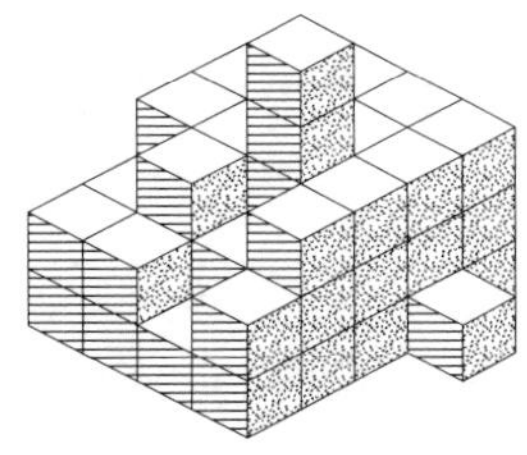

Volume: 45 cubic units
Surface area: 100 square units
Base perimeter: 18 units

45 B

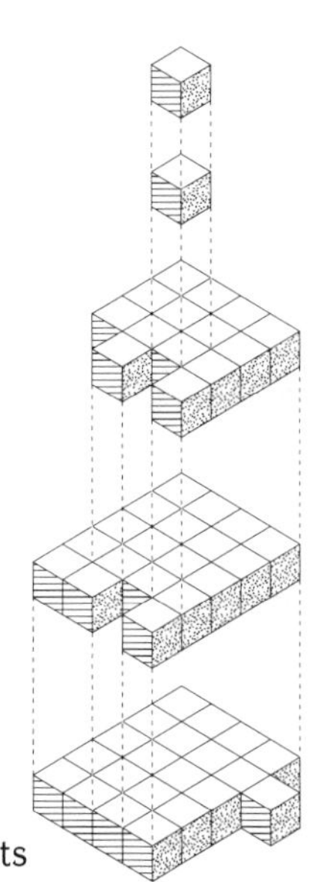

Volume: 56 cubic units
Surface area: 108 square units
Base perimeter: 20 units

Thinkcard 46

46 A

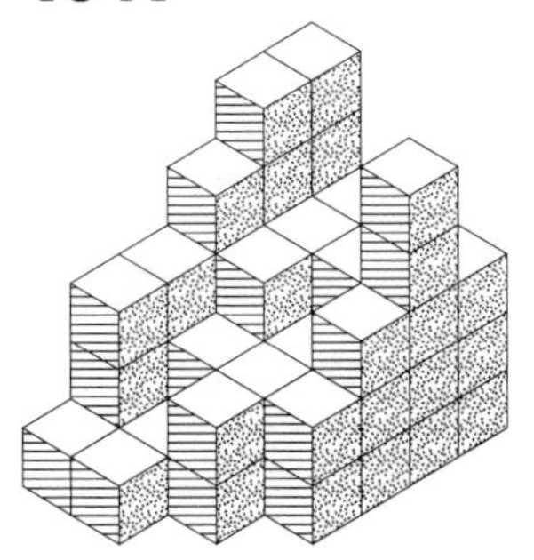

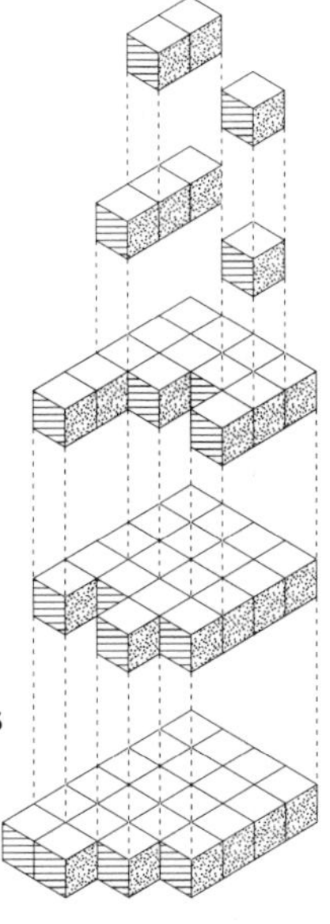

Volume: 59 cubic units
Surface area: 124 square units
Base perimeter: 20 units

46 B

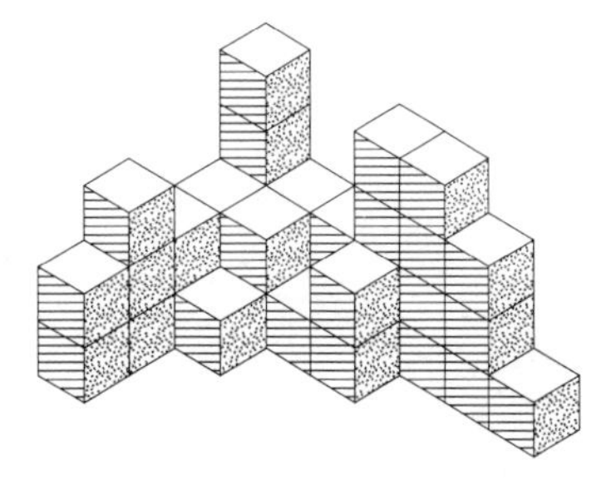

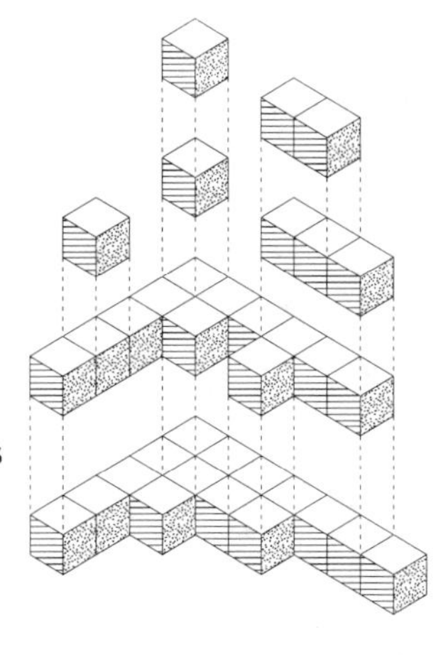

Volume: 35 cubic units
Surface area: 104 square units
Base perimeter: 24 units

Thinkcard 37

Please find the volume, total surface area, and base perimeter of these figures. The back sides are completely smooth, permitting you to infer the number of blocks in the figure. Make an exploded view isometric drawing of each figure.

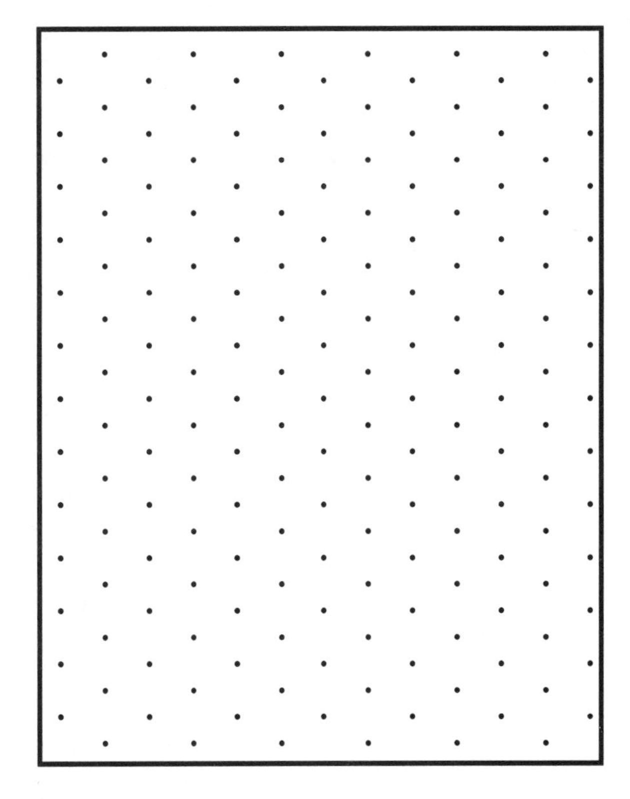

Figure A

Volume: __________

Surface area: __________

Base perimeter: __________

Figure B

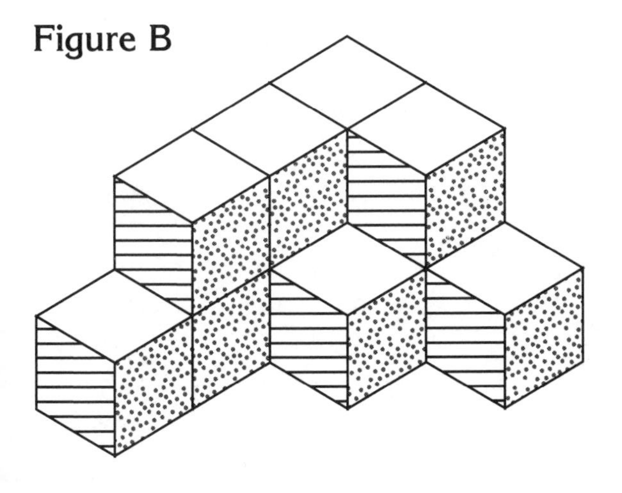

Volume: __________

Surface area: __________

Base perimeter: __________

Thinkcard 38

Please find the volume, total surface area, and base perimeter of these figures. The back sides are completely smooth, permitting you to infer the number of blocks in the figure. Make an exploded view isometric drawing of each figure.

Figure A

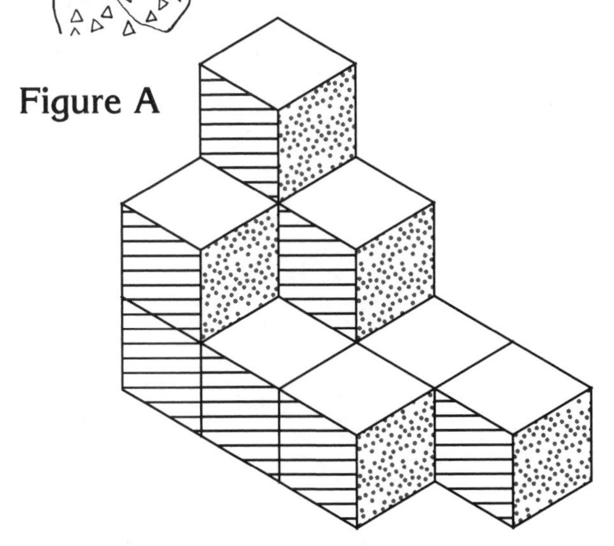

Volume: __________

Surface area: __________

Base perimeter: __________

Figure B

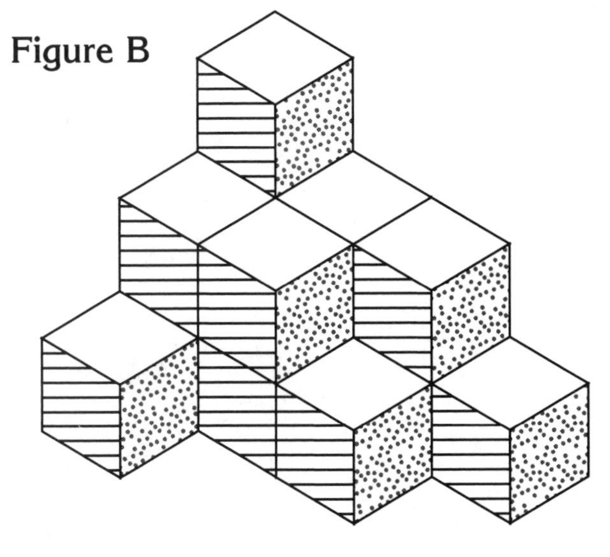

Volume: __________

Surface area: __________

Base perimeter: __________

Please find the volume, total surface area, and base perimeter of these figures. The back sides are completely smooth, permitting you to infer the number of blocks in the figure. Make an exploded view isometric drawing of each figure.

Figure A

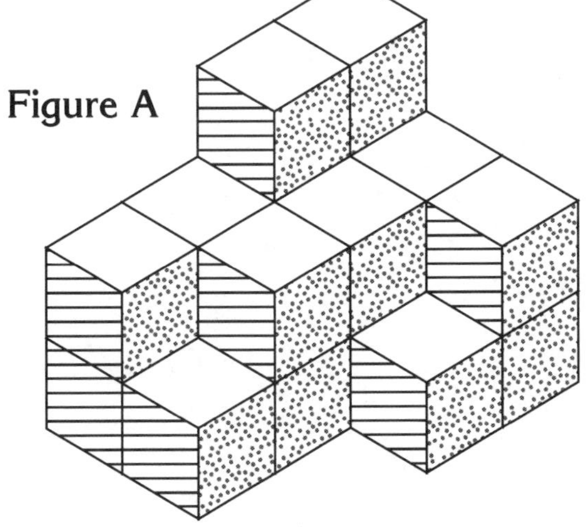

Volume: __________

Surface area: __________

Base perimeter: __________

Figure B

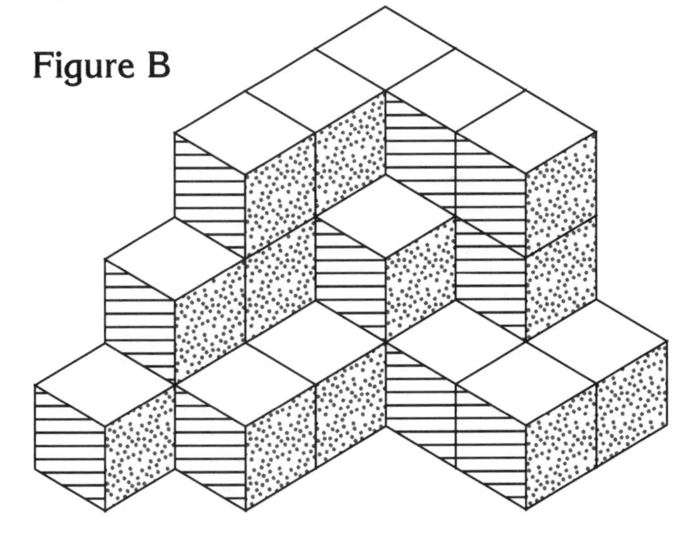

Volume: __________

Surface area: __________

Base perimeter: __________

Please find the volume, total surface area, and base perimeter of these figures. The back sides are completely smooth, permitting you to infer the number of blocks in the figure. Make an exploded view isometric drawing of each figure.

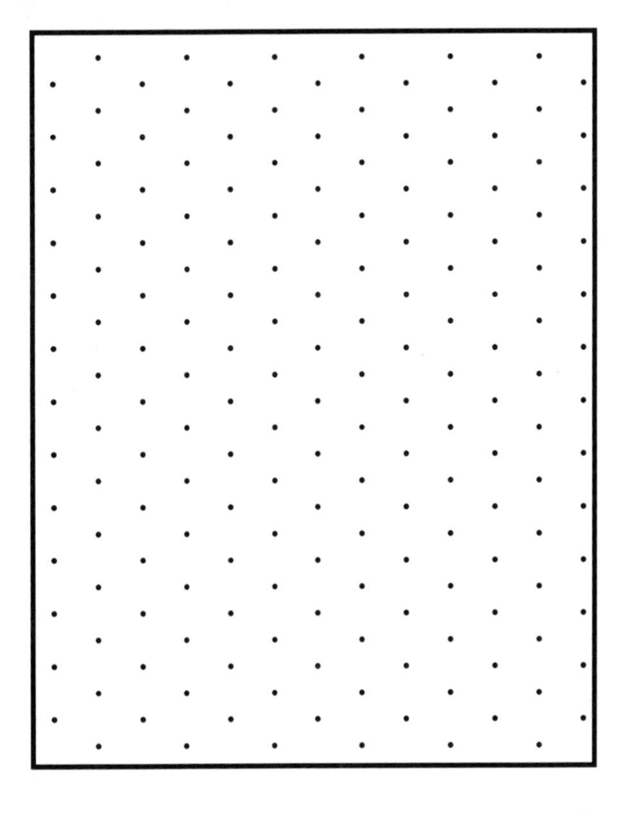

Figure A

Volume: __________

Surface area: __________

Base perimeter: __________

Figure B

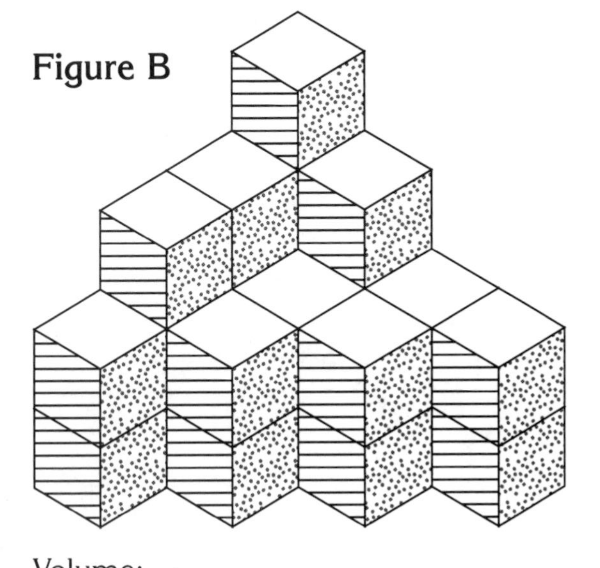

Volume: __________

Surface area: __________

Base perimeter: __________

Thinkcard 41

Please find the volume, total surface area, and base perimeter of these figures. The back sides are completely smooth, permitting you to infer the number of blocks in the figure. Make an exploded view isometric drawing of each figure.

Figure A

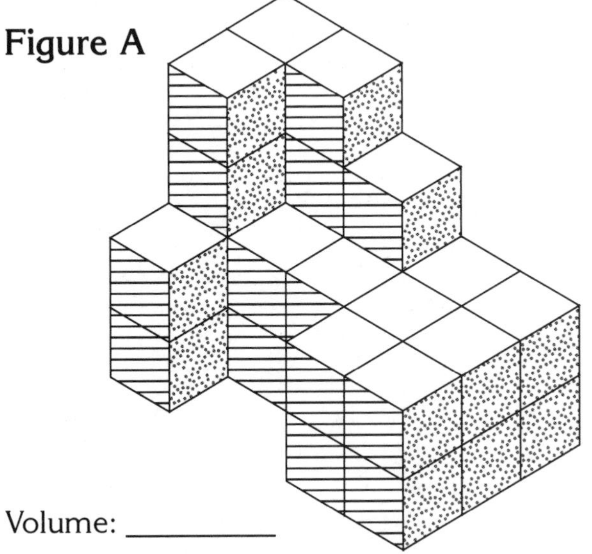

Volume: __________

Surface area: __________

Base perimeter: __________

Figure B

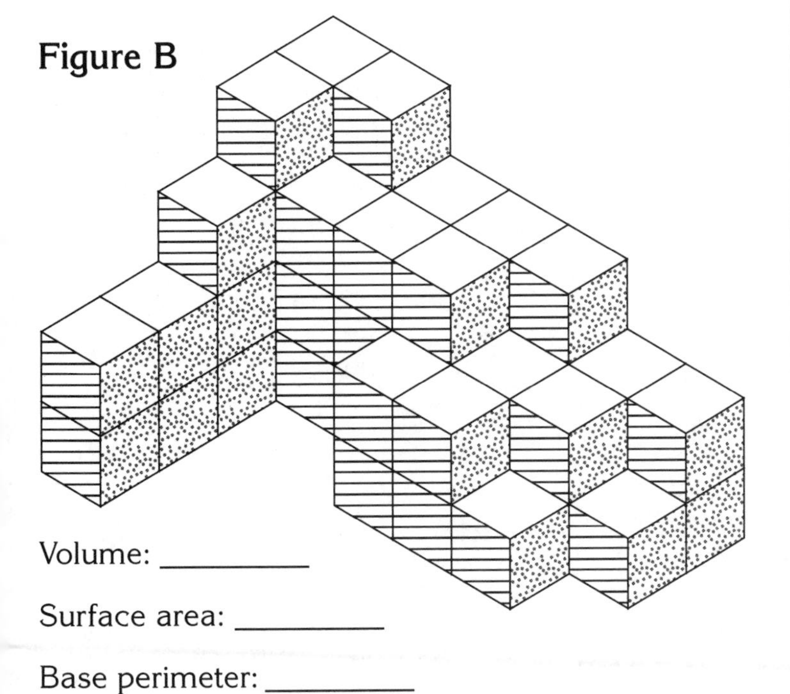

Volume: __________

Surface area: __________

Base perimeter: __________

Thinkcard 42

Please find the volume, total surface area, and base perimeter of these figures. The back sides are completely smooth, permitting you to infer the number of blocks in the figure. Make an exploded view isometric drawing of each figure.

Figure A

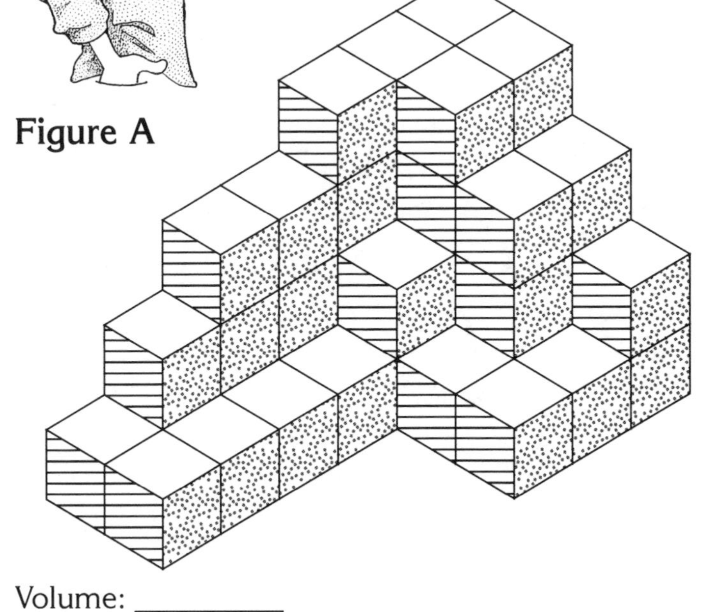

Volume: __________

Surface area: __________

Base perimeter: __________

Figure B

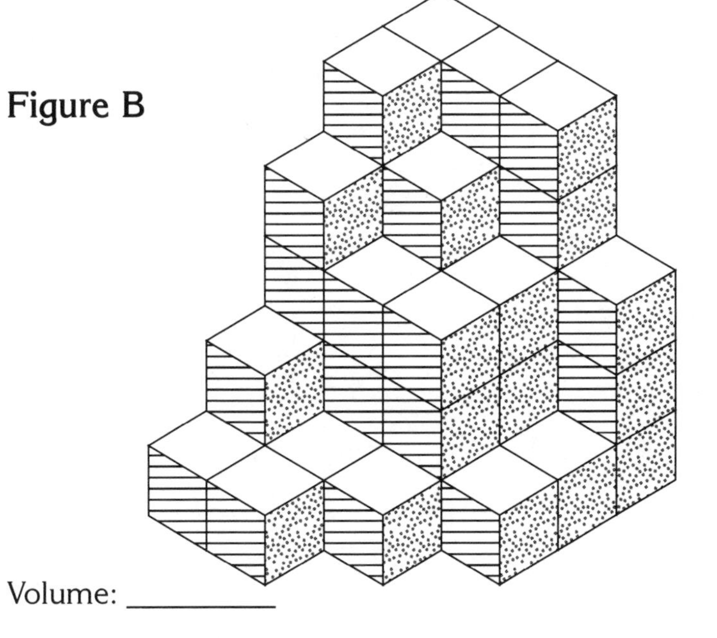

Volume: __________

Surface area: __________

Base perimeter: __________

Thinkcard 43

Please find the volume, total surface area, and base perimeter of these figures. The back sides are completely smooth, permitting you to infer the number of blocks in the figure. Make an exploded view isometric drawing of each figure.

Figure A

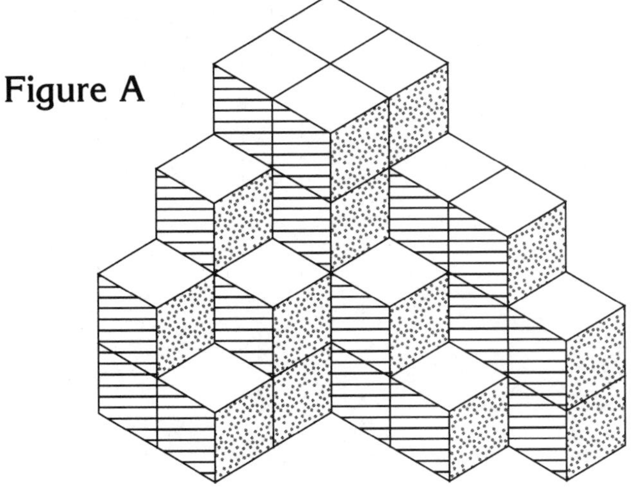

Volume: ________

Surface area: ________

Base perimeter: ________

Figure B

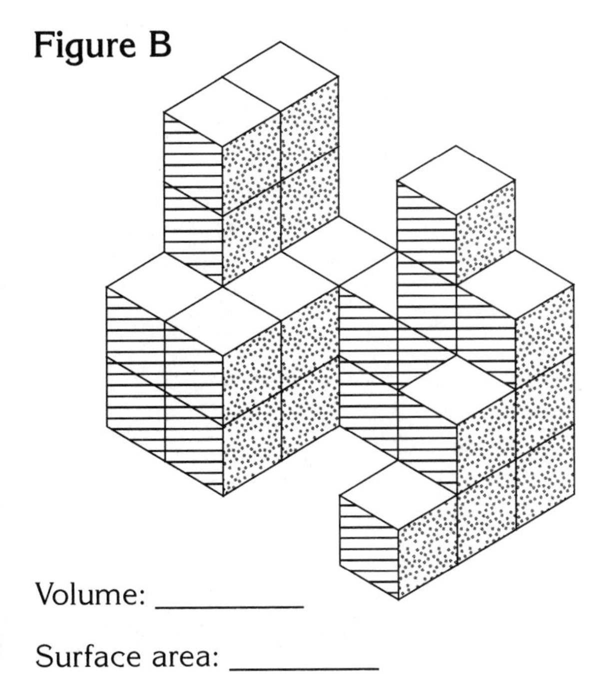

Volume: ________

Surface area: ________

Base perimeter: ________

Please find the volume, total surface area, and base perimeter of these figures. The back sides are completely smooth, permitting you to infer the number of blocks in the figure. Make an exploded view isometric drawing of each figure.

Figure A

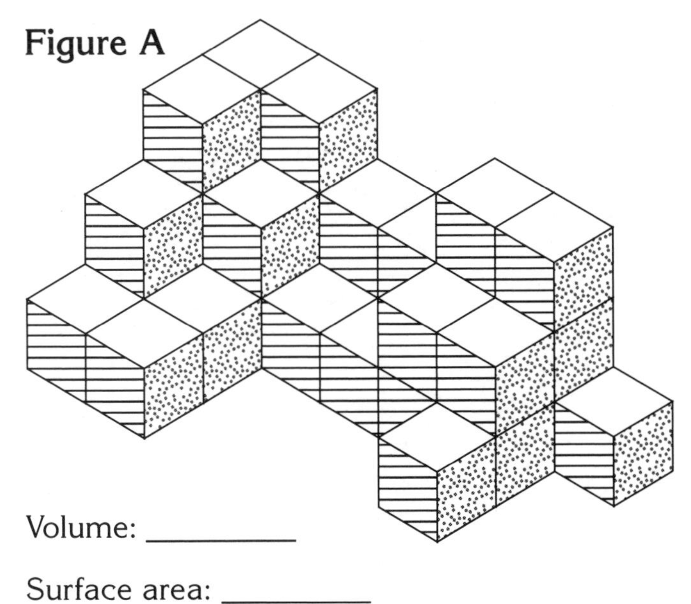

Volume: __________

Surface area: __________

Base perimeter: __________

Figure B

Volume: __________

Surface area: __________

Base perimeter: __________

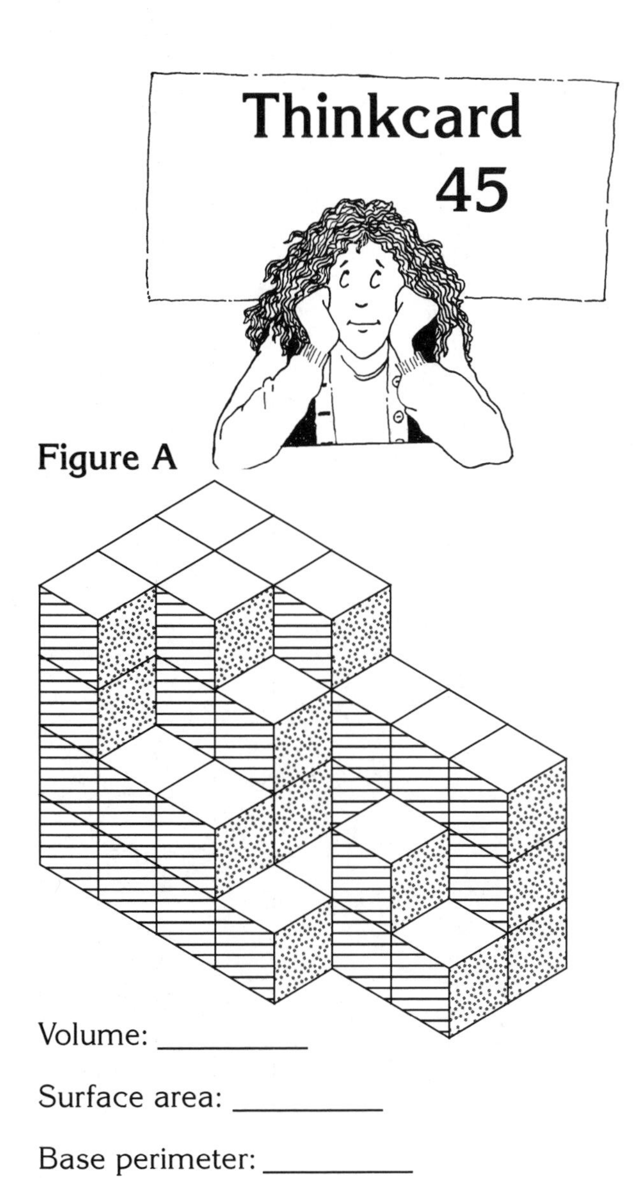

Please find the volume, total surface area, and base perimeter of these figures. The back sides are completely smooth, permitting you to infer the number of blocks in the figure. Make an exploded view isometric drawing of each figure.

Figure A

Volume: __________

Surface area: __________

Base perimeter: __________

Figure B

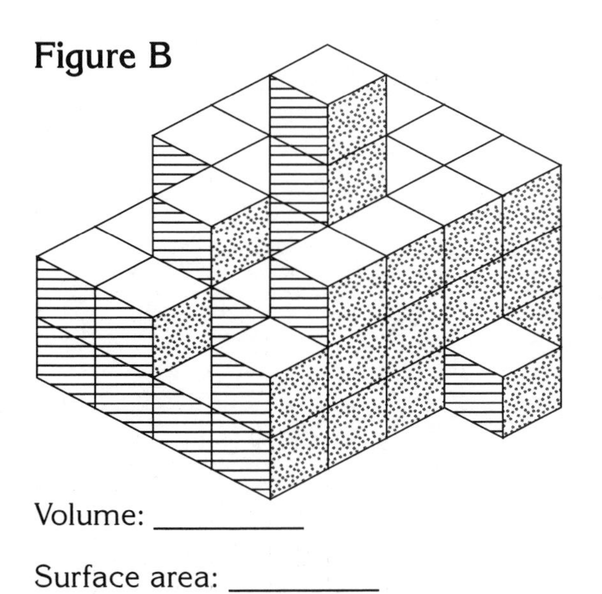

Volume: __________

Surface area: __________

Base perimeter: __________

Thinkcard 46

Please find the volume, total surface area, and base perimeter of these figures. The back sides are completely smooth, permitting you to infer the number of blocks in the figure. Make an exploded view isometric drawing of each figure.

Figure A

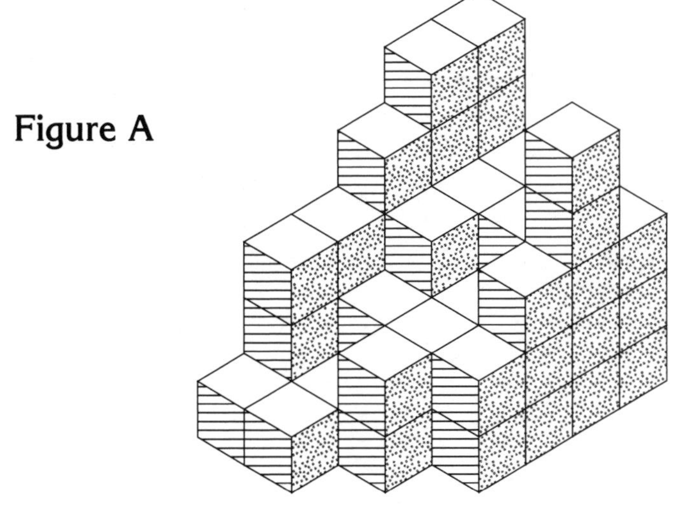

Volume: ________

Surface area: ________

Base perimeter: ________

Figure B

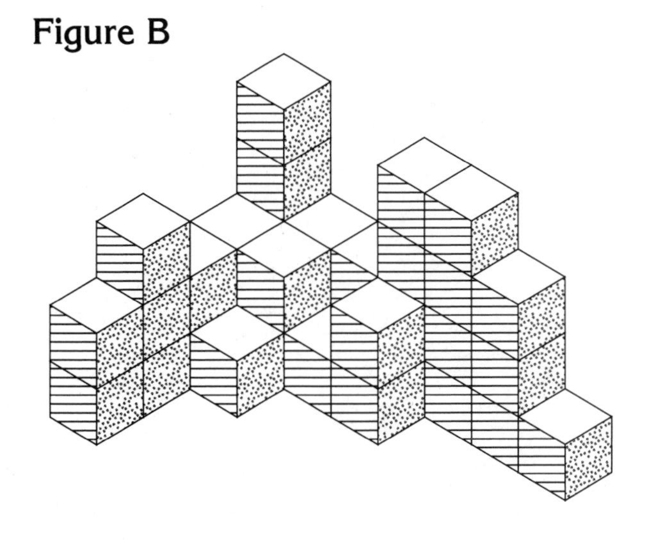

Volume: ________

Surface area: ________

Base perimeter: ________

Spatial Visualization

Part Six: Independent Investigations

Part Six provides the opportunity to engage in independent investigations using various arrangements of interlocking blocks. These are just a sample of the many opportunities that exist for using spatial visualization in context. The specific nature of each exploration, together with sample solutions where appropriate, is discussed below.

Answers

Thinkcards 47

The concepts of maximum and minimum surface areas are of great significance in mathematics. They define the outer limits of possible solutions. Hopefully, students will generalize that the minimum surface area is obtained when the dimensions of the solid are as nearly alike as possible and the maximum when the solid is as elongated as possible.

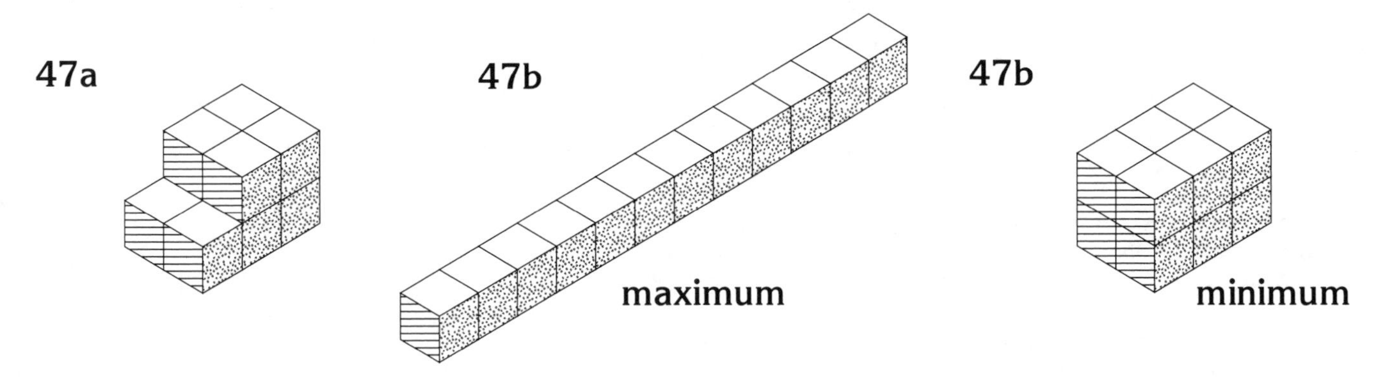

Thinkcard 48

In this activity students will construct as many shapes using eight blocks as have different total surface areas. The maximum will be a 1 x 1 x 8 solid with a surface area of 34 square units and the minimum will be a 2 x 2 x 2 solid with a surface area of 24 square units. The following surface areas are possible: 24, 28, 30, 32, and 34 square units.

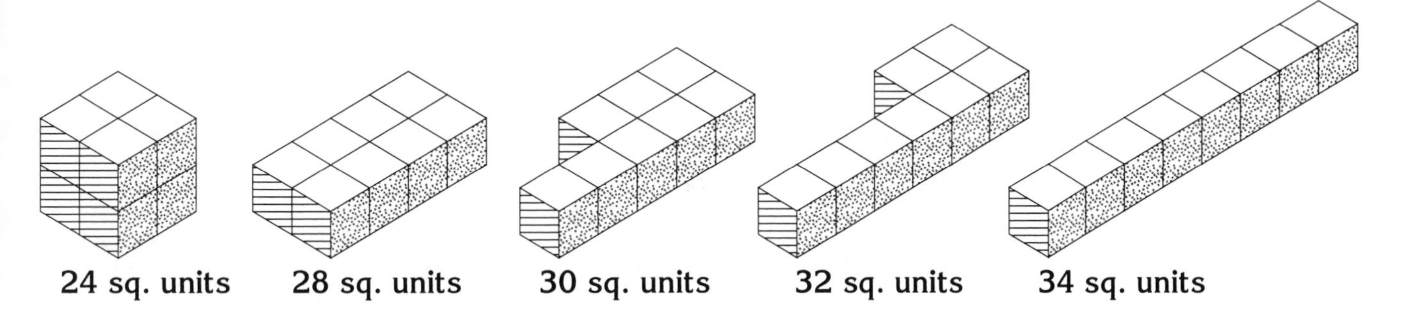

Answers

Thinkcard 49

Figures constructed with four blocks are known as *order-4 polycubes*. This activity is designed as an independent investigation rather than as a formal, full-class activity. However, it is so flexible that it lends itself to almost any use designed by the teacher. Isometric drawings enable students to keep a record of discovered shapes.

Order-4 Polycubes

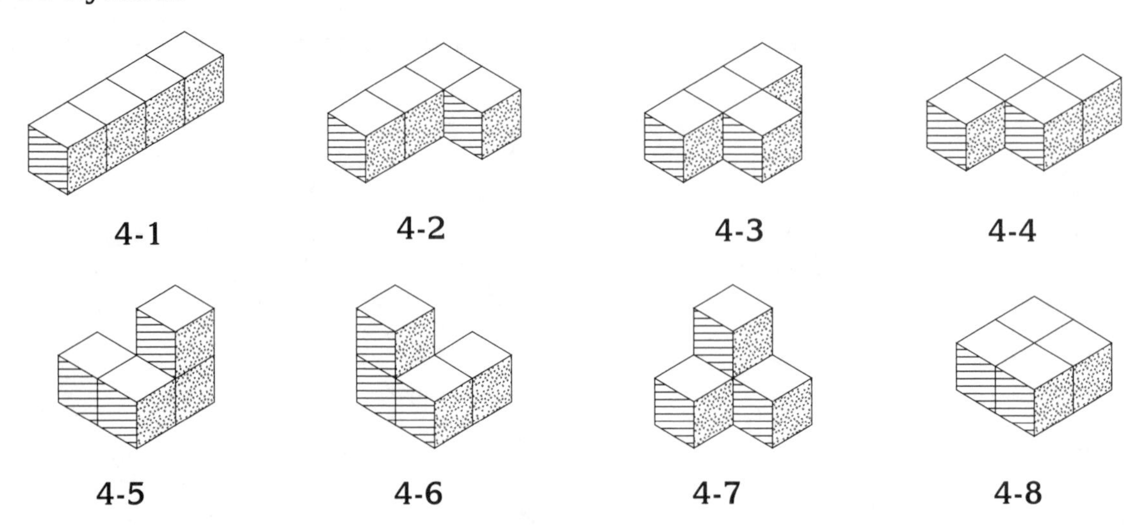

Thinkcard 50

The models in this activity belong to the family of *order-5 polycubes*. The 12 one-layer, order-5 polycubes are known as *pentominoes*. This is the same general type of exploration as in *Thinkcard 49*. Students need to be alerted to the fact that reflections or rotations are not considered different solutions. (The models resemble the letters F, L, I, P, N, T, U, V, W, X, Y, Z.)

Order-5 Single-layer Polycubes

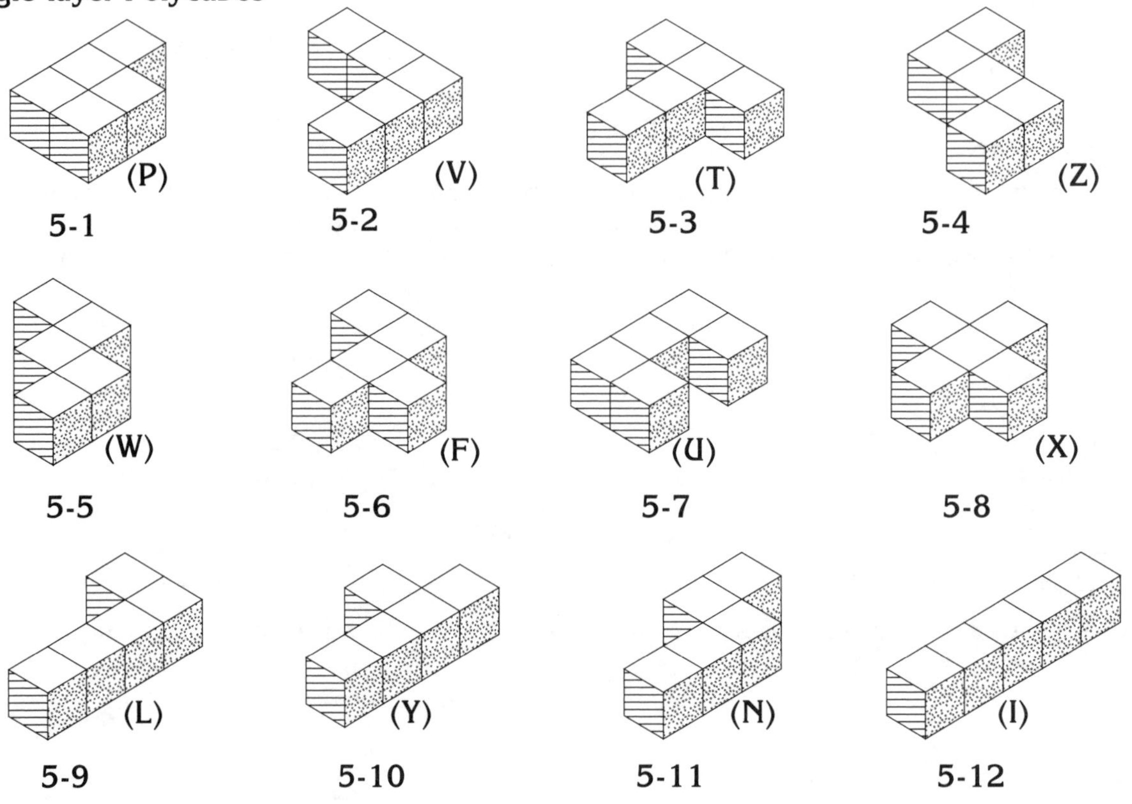

Answers

Thinkcard 51

In this exploration students try to discover all of the multi-layered figures belonging to the order-5 polycubes. Together with the single-layer figures in the previous activity, they form the full set of 29 order-5 polycubes.

Order-5 Multi-layer Polycubes

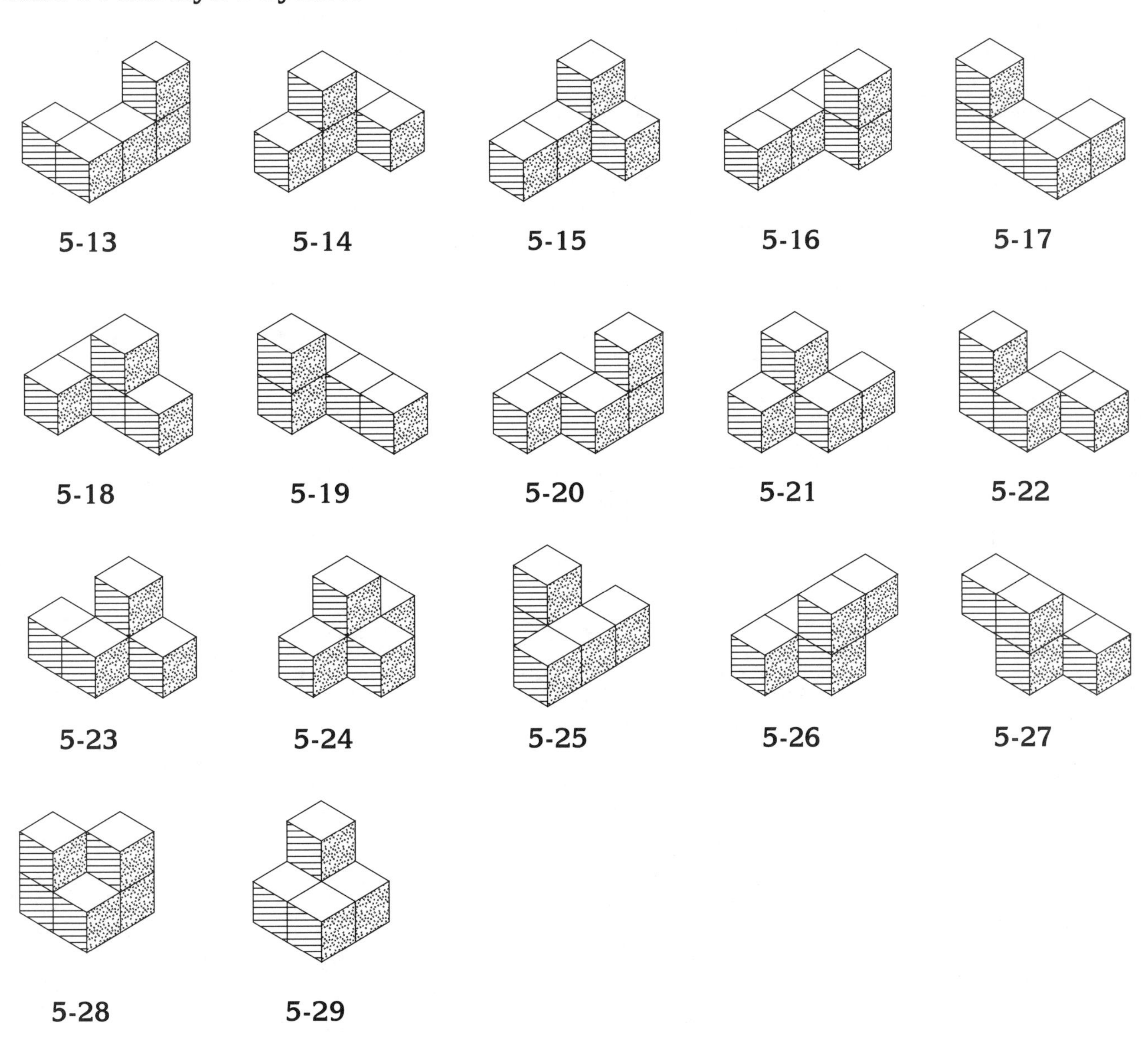

Thinkcard 52

This activity focuses on congruent solids that can be constructed using two pentominoes. The challenge may take several forms: find as many congruent pairs as possible or find as many congruent triples as possible. By making isometric drawings, students can determine that a solution is new and not one that was formed previously.

Answers

Thinkcard 53

In this activity the emphasis is on similarity and spatial visualization. By using one pentomino as a model and any nine others as construction pieces, it is possible to construct a figure that is similar to the model with each dimension three times that of the model. The total area will be nine times that of the model. Sample solutions are shown below.

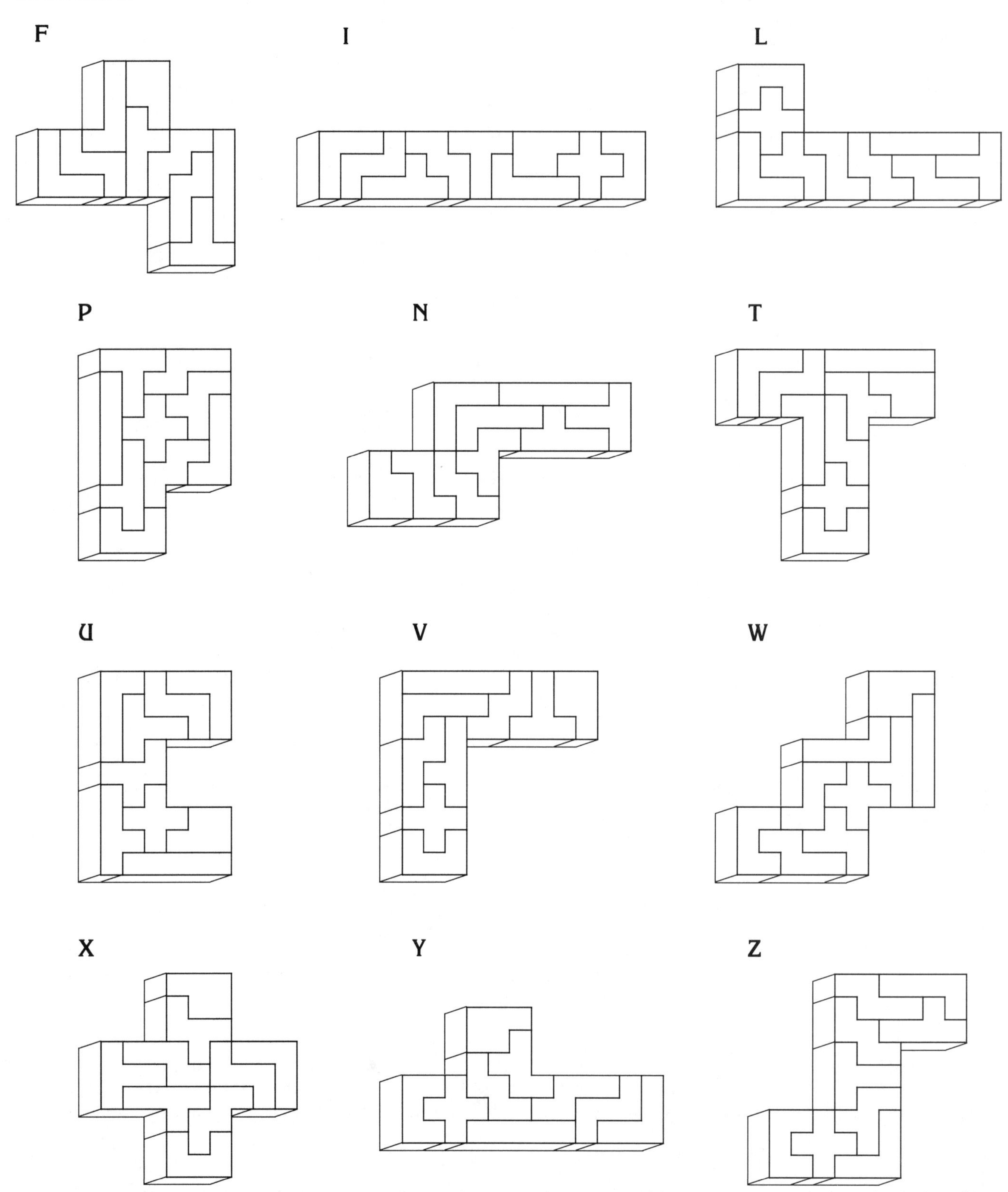

Use 10 interlocking blocks to construct a solid that has the smallest possible total surface area. Please make an isometric drawing of your solution.

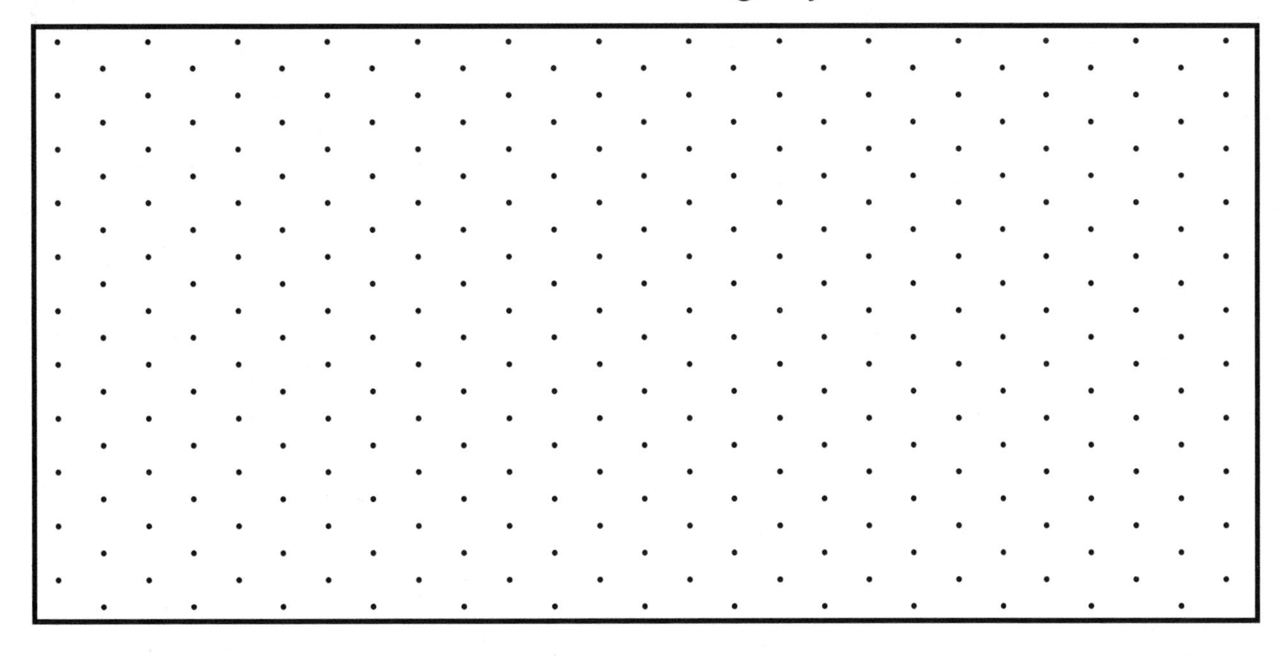

Use 12 interlocking blocks. Build the boxlike solid that has the maximum surface area and the one that has the minimum surface area. Please make an isometric drawing of each.

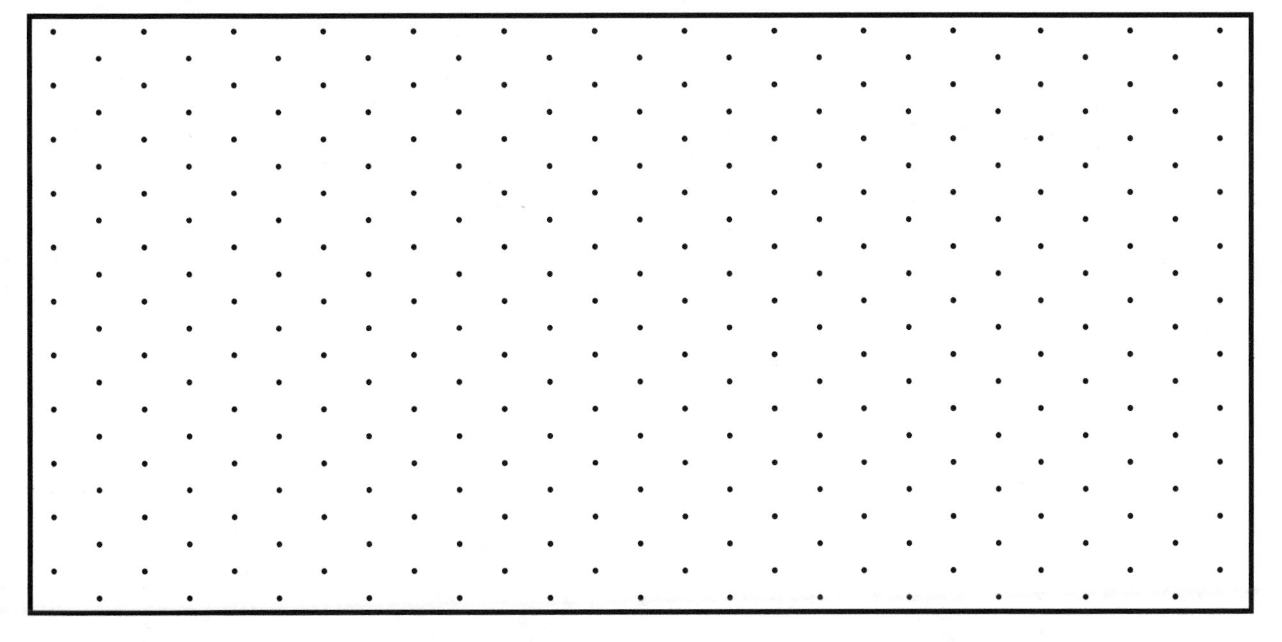

Thinkcard 48

Construct as many solids with different total surface areas as possible by using eight interlocking blocks in each. Please make isometric drawings of the solids arranged in order from the greatest to the least surface area. Label each with its surface area.

Thinkcard 49

Please find all possible figures you can construct from four interlocking blocks. The figures may have more than one layer. Please make an isometric drawing of each.(Rotations are not different solutions.)

Please find all possible one-layer figures you can construct with five interlocking blocks. For the record, please make an isometric drawing of each. (Rotations are not considered different solutions.)

Thinkcard 51

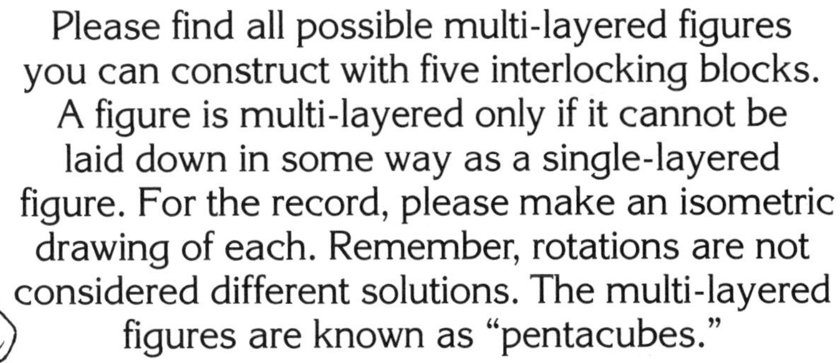

Please find all possible multi-layered figures you can construct with five interlocking blocks. A figure is multi-layered only if it cannot be laid down in some way as a single-layered figure. For the record, please make an isometric drawing of each. Remember, rotations are not considered different solutions. The multi-layered figures are known as "pentacubes."

Thinkcard 52

For this investigation use a single set of the twelve pentominoes. First, form a figure with any two pentominoes. Then, always using different pentominoes, form as many congruent figures as possible. Finally, make an isometric drawing of the largest set of congruent figures you were able to form.

Thinkcard 53

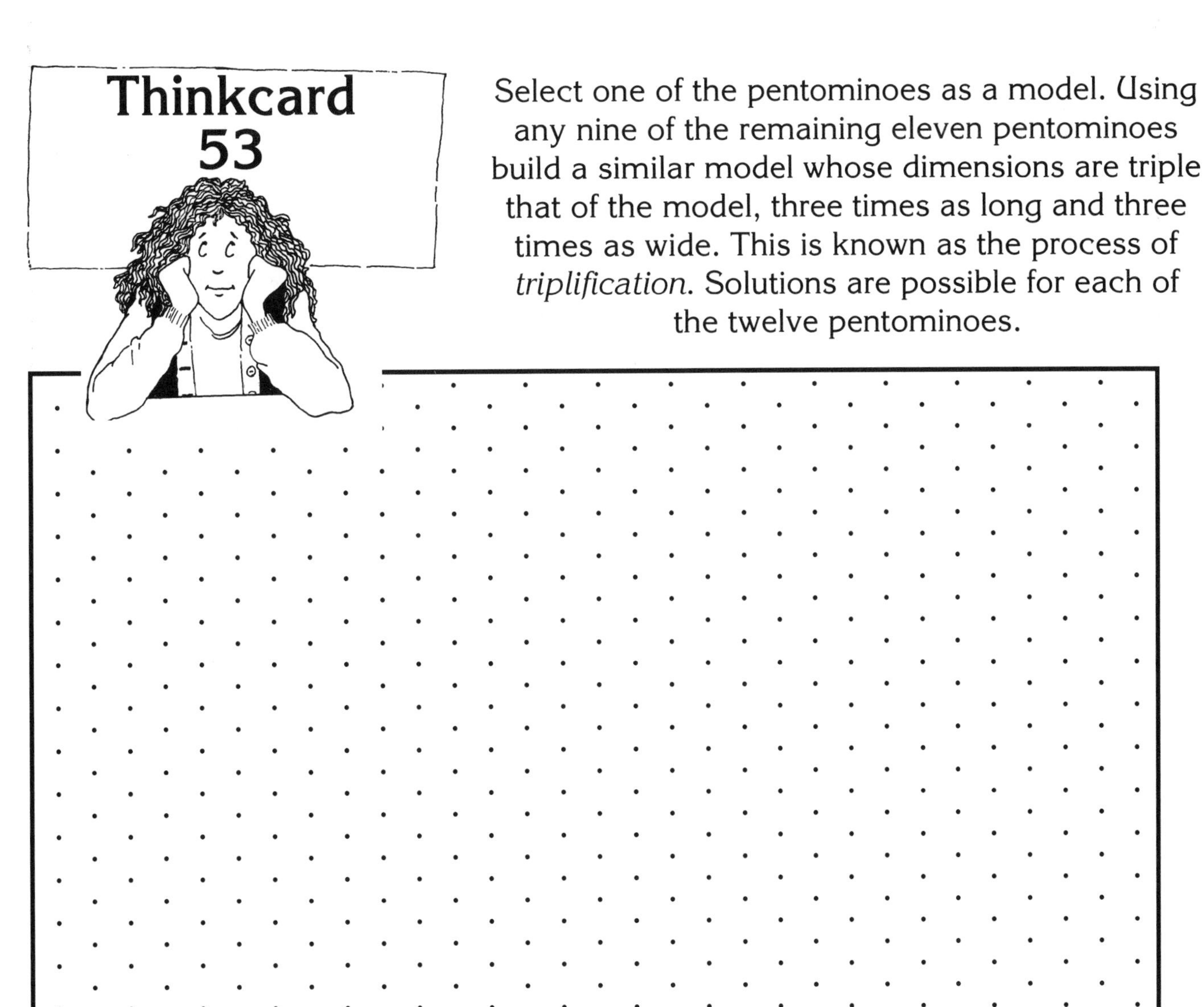

Select one of the pentominoes as a model. Using any nine of the remaining eleven pentominoes build a similar model whose dimensions are triple that of the model, three times as long and three times as wide. This is known as the process of *triplification*. Solutions are possible for each of the twelve pentominoes.

A Week with AIMS: Math Connections

Patterns, Problem Solving, and Practice Workshops

Building on 25 years of highly successful experience with the annual Fresno Pacific College "Festival of Mathematics," the AIMS program directors, research fellows, and advisors — educators with a combined total of four centuries of classroom teaching experience — are preparing a national version to be made available through a sequence of three one-week workshops. While each week is designed to stand alone, the three-year sequence provides long-term staff development. Options for continuing education during the intermediate academic years will also be made available. The general program outline is included in this article with specifics of the first year's workshop to be made available by June 1, 1997. The summer 1997 field-testing sites have been selected. Field testing will lead to further refinement before the program is released nationally in 1998.

Two Approaches to Integration

The current *A Week with AIMS: Science Connections* workshops approach the integration of science and mathematics primarily from the content of science as illustrated in this AIMS Venn diagram. The big ideas in science shape the content with mathematics serving the science.

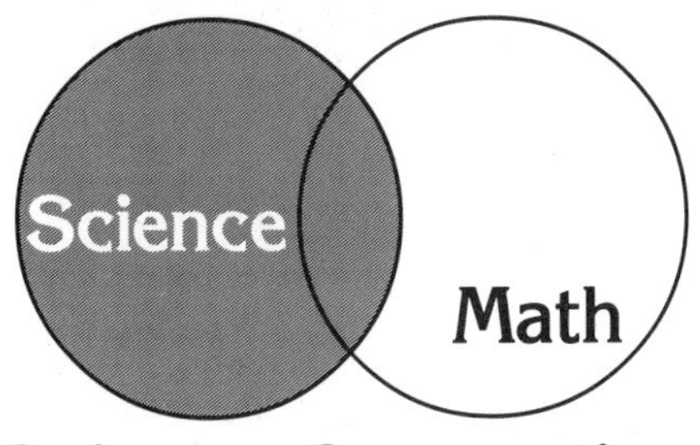

Science Connections

The new series, *A Week with AIMS: Math Connections*, approaches the integration primarily from the content of mathematics as shown in this Venn diagram. In it the big ideas of mathematics shape the content and science investigations provide the arena for application.

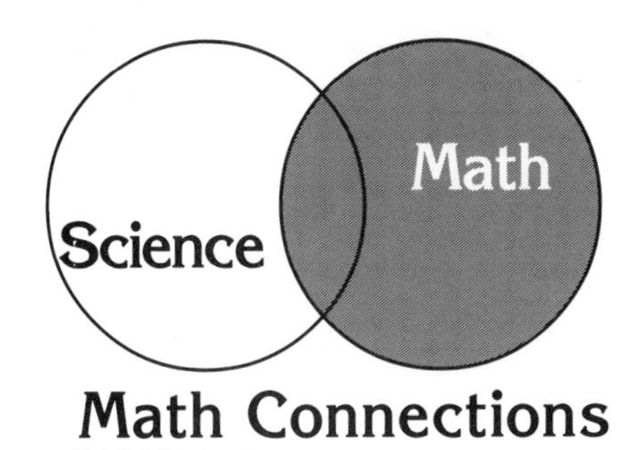

Math Connections

This dual approach recognizes and preserves the distinctive of each discipline, strengthens their integration, and moves the study of mathematics and science to a higher, more significant level. The extensive research and development resources of the AIMS Education Foundation are now focused on creating this balanced, complementary effort.

The two approaches, together with additional intermediate components, will provide multiple systems for delivering staff development on a sustained basis. AIMS has established the goal of providing the opportunity for continuous professional growth to any teacher, anywhere, at any time, with any content area of the AIMS core integrated math/science curriculum through cost-effective and time-efficient delivery systems. The delivery systems include workshops taught by AIMS facilitators, monographs, audio and video tapes, the *AIMS* magazine, Internet, etc. New developments in each will be discussed in the *AIMS* magazine.

Guidelines for Development of the New Series

The five major themes emphasized in the Grades K-4 and Grades 5-8 *Curriculum and Evaluation Standards for School Mathematics Addenda Series* of the National Council of Teachers of Mathematics provide the framework for the development of the content in this series. These themes are:

- geometry and spatial sense,
- dealing with data and chance,
- number sense and operations,
- patterns and functions, and
- understanding rational numbers and proportions.

These themes will characterize the content of the three one-week workshops and intermediate academic year activities.

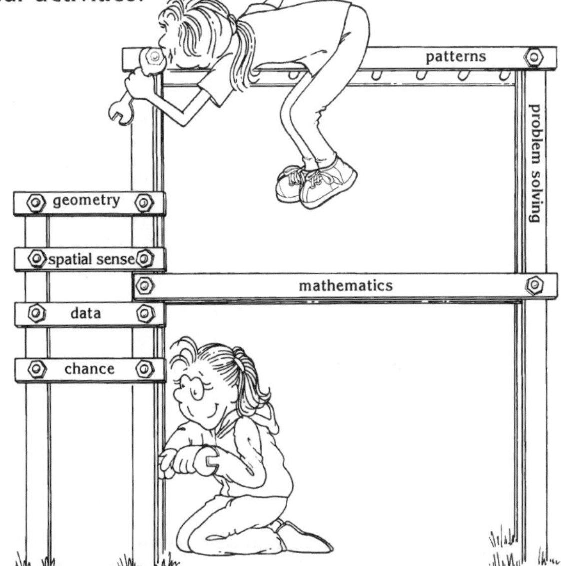

The program content is relevant for teachers anywhere in the nation. *Its foundation is competence in basic skills and its fulfillment is discovering patterns in science and mathematics in a problem-solving, investigative environment.* The pathway between is marked by abundant opportunities to construct knowledge, solve motivating and instructive puzzles, and investigate meaningful, real-life applications.

Activities are selected to be interesting so they attract students, motivating them to persist in the face of difficulties, and rewarding them so they delight in their accomplishments. Primary consideration is given to the three factors Eric Jensen suggests are critical to learner-created meaning: relevance, engaging the learner's emotions, and providing a context that relates activities to an overall pattern.

Why "Patterns, Problem Solving, and Practice"?

The AIMS curriculum development team is convinced that "Patterns, Problem Solving, and Practice" encompasses the essence of mathematics. Each of these components supports the others and together they constitute a balanced, integrated whole.

Patterns

Mathematics is the "science of patterns" and should, therefore, be approached as the study of patterns. The search for patterns should be *imbedded in every mathematical experience*, as it will be in this AIMS curriculum, rather than isolated as a special topic.

What does it mean to "do mathematics"? It is **not** performing computations. Jerry P. King describes it well: "you are doing mathematics whenever you are producing mathematics that is new to you. If you have invented the problem yourself and then solved it yourself, the work is close to what mathematicians actually do." *Students "do mathematics" when they create, search for, discover, and generalize patterns.* In this search for patterns, students have unlimited opportunity for rigorous practice with basic skills.

Students and teachers will experience the inherent interest that patterns generate and the power for solving problems they provide because all of us are pattern seekers by nature, mentally "hard-wired" to discover patterns. Eric Jensen stresses that "our brain craves meaning and the quest for it causes us to derive any pattern we can from information." Because of this fundamental human drive, students find searching for patterns to be inherently interesting and worthwhile. While the search for patterns is intriguing for gifted students, it holds equally high interest for all students. The search is valuable for uncovering the big ideas of mathematics, engaging students in doing mathematics, and helping them to make sense of the mathematical enterprise.

Readers desiring to learn more about the rationale for this AIMS approach are referred to the series entitled *Mathematics: The Science of Patterns* by Richard Thiessen beginning in the April, 1994 issue of the *AIMS* magazine and the series *Advantages of a Pattern-Based Math/Science Curriculum* by Arthur Wiebe beginning in the January, 1996 issue.

Problem Solving

AIMS defines problem-solving as "finding the solution when no obvious path to the solution exists." The solution lies beyond the mere use of algorithms and basic skills. Problem-solving students must intelligently and independently or collaboratively devise the pathway and then utilize basic skills and algorithms as needed to arrive at the solution.

Participants in these workshops will often problem solve in real-world contexts through investigations in science, by searching for and exploring number patterns, by solving meaningful puzzles, and by making mathematical discoveries.

Practice

In AIMS, practice is differentiated from drill. Drill has the connotation of endless repetition that produces a dulling effect. Practice provides for more intensive development of basic skills but in the context of interesting investigations and searches that raise the dignity level. Practice in the context of patterns is demanding, requiring complete accuracy to unlock their secrets. The reward is so motivating that students tend to persist in this strenuous mental activity.

Many of the AIMS practice activities are self-correcting, demand 100% accuracy, and can be used as independent investigations, small group, or total class activities.

History of Mathematic

Anecdotes and activities drawn from the history of mathematics will provide additional spice to the workshop experience. Participants will have the opportunity to choose from some of the extensive resources AIMS can provide in this area.

Themes Made Specific

Students will find the AIMS approach to the study of numbers to be refreshing and interesting. In geometry, the approach differs from that in common use. Studies begin with solids, the geometric objects most familiar to students. Piaget would love this! Solids embody the characteristics of volume, surface area, length of edges, angles, etc. which are abstractions. By studying solids first, students are provided a context for such abstractions that imparts meaning.

The connection between mathematics and science becomes more meaningful within the AIMS curriculum in which combinations of measurements are studied for their production of new units of measurement. For example, rate multiplied by time produces a new measure, that of distance. Mass divided by volume produces the measure of density. Surface area divided by volume gives rise to scaling, a fundamental factor for determining how often an animal eats, where it can live, how rapidly its heart beats, and how long it can be expected to live!

The general themes are outlined in the following chart:

Number Sense and Operations	Geometry and Spatial Sense	Dealing with Data and Chance	Patterns and Functions
K-2: number sense; counting; numeration; number relationships; place value; patterned arrangements; effect of operations on numbers; basic addition facts; fair shares; comparisons; mental math	K-2: observe, match, contrast, identify, describe solids; collect and use objects in child's world, investigate objects' properties and relationships; build, draw, put together, take apart, and visualize solid figures; develop understanding of length, mass, area, capacity, volume, temperature, time, and money through direct and indirect comparisons and understanding of measurement tools that will be constructed	K-2: gathering data by counting; comparing, organizing, and interpreting data; telling a story about graphs; drawing conclusions and looking at the shape of data	K-2: Patterns and functions will be integrated in all activities and will serve as the principal means for studying all of the content.
3-5: developing meaning for operations; intelligent, playful practice of basic facts/skills; alternate algorithms; place value; problem solving in mathematical microworlds; historical connections; literature correlation	3-5: sort and classify solids, observe properties; study shadows of 3-D objects, cut down solids into nets, construct solids from nets, make isometric drawings; study cubes, spheres, cylinders, cones, rectangular prisms, and pyramids; study measurable properties, use customary and non-customary units, build measuring tools, understand the approximate nature of measurement	3-5: moving from disorganization to organization; looking at relationships; learning how to select and construct tables, charts, and graphs; learning how to interpret data	3-5 Patterns and functions will be integrated in all activities, used to achieve mastery of basic skills, stimulate interest and understanding and serve as the principal means for studying all of the content.
6-9: proportional reasoning; distributive property; multiplication table patterns; multiplicative nature of proportions; common proportions; fractions, decimals, ratios, rates, percents, etc.	6-9: relationships of shapes, 2-D and 3-D dimensional studies of perimeter, area, surface area, volume, and useful ratios among them; isometric drawings; spatial visualization; study of angles, lines, rays, segments; applications in density, light, mapping, etc. with focus on combinations of measurements that produce new measurement units	6-9: decide what information for tables and graphs, how to gather and organize that information; interpret and communicate the information; study and compare numerous graph types	6-9 Patterns and functions will be integrated into all activities and utilized to build bridges of understanding among the topics of mathematics, arithmetic, and algebra. Generalization of patterns will receive major attention.

Total cost per participant is $300 for *A Week with AIMS: Math Connections.* Each participant will receive more than $120 in print and manipulative materials.

The AIMS Program

AIMS is the acronym for "**A**ctivities **I**ntegrating **M**athematics and **S**cience." Such integration enriches learning and makes it meaningful and holistic. AIMS began as a project of Fresno Pacific University to integrate the study of mathematics and science in grades K-9, but has since expanded to include language arts, social studies, and other disciplines.

AIMS is a continuing program of the non-profit AIMS Education Foundation. It had its inception in a National Science Foundation funded program whose purpose was to explore the effectiveness of integrating mathematics and science. The project directors in cooperation with 80 elementary classroom teachers devoted two years to a thorough field-testing of the results and implications of integration.

The approach met with such positive results that the decision was made to launch a program to create instructional materials incorporating this concept. Despite the fact that thoughtful educators have long recommended an integrative approach, very little appropriate material was available in 1981 when the project began. A series of writing projects have ensued and today the AIMS Education Foundation is committed to continue the creation of new integrated activities on a permanent basis.

The AIMS program is funded through the sale of this developing series of books and proceeds from the Foundation's endowment. All net income from program and products flows into a trust fund administered by the AIMS Education Foundation. Use of these funds is restricted to support of research, development, and publication of new materials. Writers donate all their rights to the Foundation to support its on-going program. No royalties are paid to the writers.

The rationale for integration lies in the fact that science, mathematics, language arts, social studies, etc., are integrally interwoven in the real world from which it follows that they should be similarly treated in the classroom where we are preparing students to live in that world. Teachers who use the AIMS program give enthusiastic endorsement to the effectiveness of this approach.

Science encompasses the art of questioning, investigating, hypothesizing, discovering, and communicating. Mathematics is the language that provides clarity, objectivity, and understanding. The language arts provide us powerful tools of communication. Many of the major contemporary societal issues stem from advancements in science and must be studied in the context of the social sciences. Therefore, it is timely that all of us take seriously a more holistic mode of educating our students. This goal motivates all who are associated with the AIMS Program. We invite you to join us in this effort.

Meaningful integration of knowledge is a major recommendation coming from the nation's professional science and mathematics associations. The American Association for the Advancement of Science in *Science for All Americans* strongly recommends the integration of mathematics, science, and technology. The National Council of Teachers of Mathematics places strong emphasis on applications of mathematics such as are found in science investigations. AIMS is fully aligned with these recommendations.

Extensive field testing of AIMS investigations confirms these beneficial results.

1. Mathematics becomes more meaningful, hence more useful, when it is applied to situations that interest students.
2. The extent to which science is studied and understood is increased, with a significant economy of time, when mathematics and science are integrated.
3. There is improved quality of learning and retention, supporting the thesis that learning which is meaningful and relevant is more effective.
4. Motivation and involvement are increased dramatically as students investigate real-world situations and participate actively in the process.

We invite you to become part of this classroom teacher movement by using an integrated approach to learning and sharing any suggestions you may have. The AIMS Program welcomes you!

AIMS Education Foundation Programs

A Day with AIMS

Intensive one-day workshops are offered to introduce educators to the philosophy and rationale of AIMS. Participants will discuss the methodology of AIMS and the strategies by which AIMS principles may be incorporated into curriculum. Each participant will take part in a variety of hands-on AIMS investigations to gain an understanding of such aspects as the scientific/mathematical content, classroom management, and connections with other curricular areas. *A Day with AIMS* workshops may be offered anywhere in the United States. Necessary supplies and take-home materials are usually included in the enrollment fee.

A Week with AIMS

Throughout the nation, AIMS offers many one-week workshops each year, usually in the summer. Each workshop lasts five days and includes at least 30 hours of AIMS hands-on instruction. Participants are grouped according to the grade level(s) in which they are interested. Instructors are members of the AIMS Instructional Leadership Network. Supplies for the activities and a generous supply of take-home materials are included in the enrollment fee. Sites are selected on the basis of applications submitted by educational organizations. If chosen to host a workshop, the host agency agrees to provide specified facilities and cooperate in the promotion of the workshop. The AIMS Education Foundation supplies workshop materials as well as the travel, housing, and meals for instructors.

AIMS One-Week Perspectives Workshops

Each summer, Fresno Pacific University offers AIMS one-week workshops on its campus in Fresno, California. AIMS Program Directors and highly qualified members of the AIMS National Leadership Network serve as instructors.

The Science Festival and the Festival of Mathematics

Each summer, Fresno Pacific University offers a Science Festival and a Festival of Mathematics. These festivals have gained national recognition as inspiring and challenging experiences, giving unique opportunities to experience hands-on mathematics and science in topical and grade-level groups. Guest faculty includes some of the nation's most highly regarded mathematics and science educators. Supplies and take-home materials are included in the enrollment fee.

The AIMS Instructional Leadership Program

This is an AIMS staff-development program seeking to prepare facilitators for leadership roles in science/math education in their home districts or regions. Upon successful completion of the program, trained facilitators become members of the AIMS Instructional Leadership Network, qualified to conduct AIMS workshops, teach AIMS in-service courses for college credit, and serve as AIMS consultants. Intensive training is provided in mathematics, science, process and thinking skills, workshop management, and other relevant topics.

College Credit and Grants

Those who participate in workshops may often qualify for college credit. If the workshop takes place on the campus of Fresno Pacific University, that institution may grant appropriate credit. If the workshop takes place off-campus, arrangements can sometimes be made for credit to be granted by another college or university. In addition, the applicant's home school district is often willing to grant in-service or professional development credit. Many educators who participate in AIMS workshops are recipients of various types of educational grants, either local or national. Nationally known foundations and funding agencies have long recognized the value of AIMS mathematics and science workshops to educators. The AIMS Education Foundation encourages educators interested in attending or hosting workshops to explore the possibilities suggested above. Although the Foundation strongly supports such interest, it reminds applicants that they have the primary responsibility for fulfilling *current* requirements.

For current information regarding the programs described above, please complete the following:

Information Request

Please send current information on the items checked:

____ *Basic Information Packet* on AIMS materials
____ *Festival of Mathematics*
____ *Science Festival*
____ *AIMS Instructional Leadership Program*
____ *AIMS One-Week Perspectives* workshops
____ *A Week with AIMS* workshops
____ Hosting information for *A Day with AIMS* workshops
____ Hosting information for *A Week with AIMS* workshops

Name ______________________________ Phone ______________

Address ____
Street City State Zip

We invite you to subscribe AIMS!